SpringerBriefs in Food, Health, and Nutrition

For further volumes:
http://www.springer.com/series/10203

Jana Žel · Mojca Milavec · Dany Morisset
Damien Plan · Guy Van den Eede
Kristina Gruden

How to Reliably Test for GMOs

 Springer

Jana Žel
National Institute of Biology
Večna pot 111
1000 Ljubljana
Slovenia
Jana.Zel@nib.si

Dany Morisset
National Institute of Biology
Večna pot 111
1000 Ljubljana
Slovenia
Dany.Morisset@nib.si

Guy Van den Eede
European Commission
Joint Research Centre
Institute for Health and Consumer
Protection Via E. Fermi 2749
21027 Ispra (Varese)
Italy
Guy.VAN-DEN-EEDE@ec.europa.eu

Mojca Milavec
National Institute of Biology
Večna pot 111
1000 Ljubljana
Slovenia
Mojca.Milavec@nib.si

Damien Plan
European Commission
Joint Research Centre
Institute for Health and Consumer
Protection Via E. Fermi 2749
21027 Ispra (Varese)
Italy
Damien.PLAN@ec.europa.eu

Kristina Gruden
National Institute of Biology
Večna pot 111
1000 Ljubljana
Slovenia
Kristina.Gruden@nib.si

ISBN 978-1-4614-1389-9   e-ISBN 978-1-4614-1390-5
DOI 10.1007/978-1-4614-1390-5
Springer New York Dordrecht Heidelberg London

Library of Congress Control Number: 2011936792

Printed on acid-free paper

Springer is part of Springer Science+Business Media (www.springer.com)

Preface

Testing of genetically modified organisms (GMOs) presents significant challenges to the laboratories assigned to this specific task. It requires, among other skills, a profound understanding of molecular biology, the capacity to set up and to use accurate procedures and methods of analyses, including quality management systems, and a permanent adaptation to and knowledge of new GMOs entering the global market.

This book reflects the practical experience and knowledge gained over many years of various activities at the National Institute of Biology (NIB), Slovenia, which is a research organization and National Reference Laboratory (NRL) as well as a laboratory performing routine analyses of food, feed, and seeds. It operates in tight collaboration with the European Network of GMO Laboratories (ENGL) and the Institute for Health and Consumer Protection (IHCP) of the European Commission (EC) Joint Research Centre (JRC). This book is written in cooperation with the IHCP which has extensive knowledge and experience regarding GMO detection and which hosts the European Union Reference Laboratory for GM Food and Feed (EU-RL GMFF) and chairs the ENGL. In addition, high-quality guidance documents produced by various ENGL working groups were of valuable support in writing this book.

This book is intended as an aid to the authorities and testing laboratories, giving essential information on GMO legislation and testing and in addition, profound and precise practical information about the implementation of real-time PCR for both qualitative and quantitative analysis.

PCR-based methods are the methods of choice in the European Union for GMO testing and are becoming recognized as the standards for reference methods. Legal thresholds were set for labeling GMO presence in food and feed in different countries, stimulating the development of approaches for precise quantification of DNA.

Validation and verification of the laboratory methods are two of the prerequisites for quality-assured GMO testing. The parameters of the methods that need to be evaluated, their acceptance criteria, and performance requirements are described in this book.

Measurement uncertainty can significantly influence the decision-making process; therefore it is necessary to set harmonized approaches for measurement uncertainty estimation in order to avoid international disputes caused by differences in the interpretation of results.

A glance at the pipeline of GMOs currently in development suggests that many more and diverse organisms will need to be detected in the future. Moreover, new technologies are being introduced for the further modification of organisms. This diversity in terms of new species, new traits, and new types of modified organisms presents real challenges for future detection. On the other hand, new detection methodologies and techniques are in development, offering high throughput, cost-efficient, reliable, and accurate analysis. The combination of laboratory solutions with bioinformatics tools is expected to be a successful key and approach to meeting new challenges.

There is always room for improvement in the domain of GMO analysis and each of the topics described in the book is constantly evolving. In each section of the book, references to the most informative, comprehensive, and recent literature or websites are given to offer the reader additional information.

The methodological approaches described in the present document are also relevant for other areas where detection and identification rely on nucleic acid-based methods. Additionally, this book can be used by lecturers looking for information about nucleic acid detection and quantification. Metrological topics presented, such as validation, verification of methods, and measurement uncertainty, as well as solutions to guarantee quality assurance, can be of additional importance for the experts in laboratories dealing regularly with implementation of new methods and the setting of laboratory conditions to obtain accurate test results.

Ljubljana, Slovenia	Jana Žel
Ljubljana, Slovenia	Mojca Milavec
Ljubljana, Slovenia	Dany Morisset
Ispra, Italy	Damien Plan
Ispra, Italy	Guy Van den Eede
Ljubljana, Slovenia	Kristina Gruden

Acknowledgements

We would like to show our great gratitude to colleagues at the National Institute of Biology, especially the statistical consultant Dr. Andrej Blejec for his patient explanations and ideas in measurement uncertainty, the quality manager Dr. Marjana Camloh for useful suggestions about the quality system section, Dejan Štebih and Tina Demšar for fruitful discussions on practical laboratory views in GMO testing, and Dr. Katja Cankar and Dr. Tanja Dreo for their valuable contribution in setting up the GMO detection system at the National Institute of Biology.

We thank Steven Price for his corrections of English, Dr. Zdravko Podlesek for kindly drawing the figures, and Prof. Dr. Marina Dermastia and Dr. Aleš Kladnik for the figure on maize tissue ploidy.

The authors co-operated in many working groups and meetings of the ENGL and various research projects and would like to thank colleagues worldwide for their valuable discussions over many years of our common establishment of the GMO testing system in the European Union which also influenced this book.

This book was reviewed by Dr. Lotte Hough from the Laboratory of Diagnostics in Plants, Seed, and Feed Department of Plants and Plant Health, The Danish Plant Directorate, to whom we would like to express our sincere gratitude for her valuable suggestions.

The work of Prof. Dr. Jana Žel, Dr. Mojca Milavec, Dr. Dany Morisset, and Prof. Dr. Kristina Gruden is supported by the Slovenian Research Agency (contract no P4-0165) and Slovenian Ministry of Environment and spatial planning (contract no. 2511-10-200006).

Jana Žel
Mojca Milavec
Dany Morisset
Damien Plan
Guy Van den Eede
Kristina Gruden

Contents

How to Reliably Test for GMOs

Introduction

The last decade has seen the intensive development of GMO detection methods, additionally stimulated by the establishment of legislation frames and requests for traceability and labeling in many countries, worldwide. The GMO content thresholds set for GMO labeling have emphasized the importance of the optimization of quantitative analyses, especially using real-time PCR (qPCR) technology. The methodology of detection itself, as well as the whole procedure including the validation and verification of methods, measurement uncertainty, and the interpretation of results, were established, thus resulting in a well-performing testing system to support GMO traceability.

GMO development is a fast-growing domain and numerous new events are in the pipeline for commercialization in the coming years (Stein and Rodriguez-Cerezo 2009a, b). Moreover, these new biotech organisms will harbor novel and sometimes unique characteristics (Stein and Rodriguez-Cerezo 2009a). On the other hand, a sustained effort exists to develop and optimize efficient, cheaper, and high-throughput detection methodologies and technologies to cope with all the new challenges posed by the diverse and growing number of GM events to be analyzed.

Genetically Modified Organisms

Biotechnology and in particular recombinant deoxyribonucleic acid (rDNA) technology can be used to produce genetic modifications resulting in organisms with traits that are not present in their unmodified relatives. The definitions of such GMOs can vary in different countries, and also other terms, such as "biotech crops" for GM or transgenic plants are used (see the section, Legislation on GMO Labeling

J. Žel et al., *How to Reliably Test for GMOs*, SpringerBriefs in Food, Health, and Nutrition, DOI 10.1007/978-1-4614-1390-5_1, © Jana Žel, Mojca Milavec, Dany Morisset, Damien Plan, Guy Van den Eede, Kristina Gruden 2012

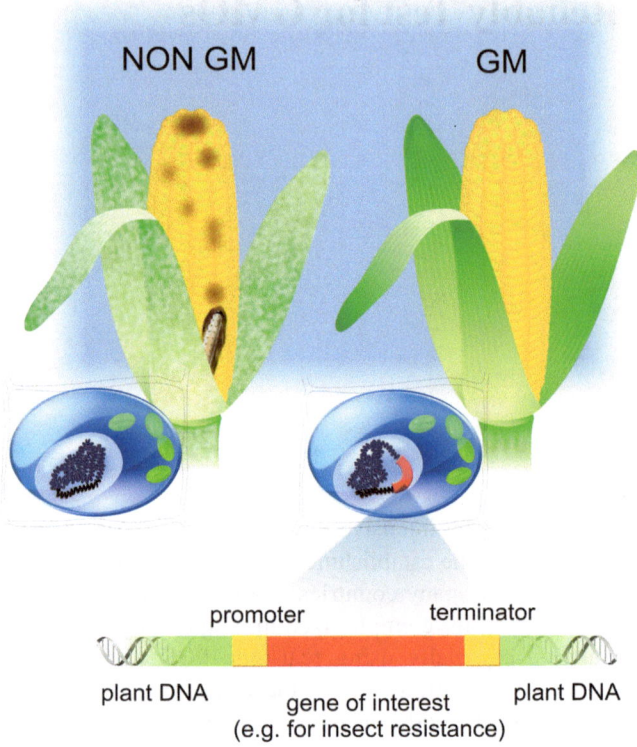

Fig. 1 Comparison of GM insect resistant and non-GM maize. The recombinant DNA sequence present in GM maize, target for the DNA-based detection of GM, is enlarged

and GMO Detection). Genetic sequence coding for proteins that result in diverse traits can be introduced in micro-organisms, plants, and animals. Most of the recent GMOs released on the world market are GM plants. Genetic sequences are introduced in plants by different methods, most commonly by *Agrobacterium*-mediated transformation or by the biolistic particle delivery system (Barampuram and Zhang 2011). This introduced rDNA sequence, also called a transgenic sequence, which differentiates the GM plant from its non-GM counterpart at the DNA level, is used as a target for the DNA-based detection of the GM plants (Fig. 1). The rDNA sequence is introduced in a unique place in the plant genome resulting in a so-called GM event. Recombinant proteins produced by the GM plant can also be used as detection targets if using immunological methods, although protein-based detection methods present some limitations compared to DNA-based detection methods (see the section, Methods).

Presence of GMOs on the World Market

GMOs are widespread in many parts of the world, their species and trait diversity rising each year (James 2011). Recent publications provide the global status of GM crops currently in commercial use, research, and development (Dymond and Hurr 2010; Stein and Rodriguez-Cerezo 2010).

Soya bean continues to be the principal biotech crop in 2010 with 81% of all soya beans grown being GM, followed by GM maize, cotton, and canola (James 2011). Herbicide tolerance is the most dominant trait deployed in soya bean, maize, canola, cotton, sugar beet, and alfalfa, followed by stacked herbicide resistance and insect-resistance traits and finally those GMs with only insect resistance (James 2011). There are also some other GM plants approved for commercial use, including virus-resistant papaya, plum and squash, insect-resistant potato and potato with altered starch composition, herbicide-tolerant rice, and carnations (http://cera-gmc. org/index.php?action=gm_crop_database).

Some vegetable plants are close to commercialization, such as insect-resistant eggplant in India, and others are in development. Most of the events in development are still herbicide-tolerant and/or insect-resistant traits. More than half are varieties of the four major crops: maize, soya bean, oilseed rape, and cotton. New crops and traits are also being introduced. Fourteen new species are well advanced in regulatory, research, or development pipelines, including rice, eucalypt, blue rose, potato, eggplant, tomato, sorghum, sugarcane, and beans. GM wheat is in early-phase field trials, with commercialization expected in 2017, and GM forage species such as clover, ryegrasses, and tall fescues, may also be expected (Dymond and Hurr 2010). New traits introduced are often related to tolerance towards greater extremes of climate and/or improvements in nutrient supply (such as drought tolerance and nitrogen-use efficiency; Dymond and Hurr 2010). Crops with more than one GM trait are progressively replacing single-trait crops. A quadruple-stacked maize was approved last year (Genuity SmartStax™) and an octet hybrid is expected in the next few years (Dymond and Hurr 2010).

Many research and development studies on GMOs are underway not only in plants, but also in other organisms thus increasing the complexity of their control and detection (see the section, New Challenges).

Information on GMOs and Their Detection on the Web

A list of websites containing information about GMOs, proficiency tests running in the area of GMO testing, and available reference materials and methods databases are provided in Table 1 as sources of information relevant for GMO testing.

Table 1 List of websites useful for detection of GMOs

Short description	Website
Sources of Information on GMOs	
The Biosafety Clearing-House (BCH) is a mechanism setup by the Cartagena Protocol on Biosafety to facilitate the exchange of information on Living Modified Organisms (LMOs) and assist the parties to better comply with their obligations under the Protocol. It provides global access to a variety of scientific, technical, environmental, legal, and capacity-building information and links to National BCH.	http://bch.cbd.int/
BCH also includes LMO Registry.	http://bch.cbd.int/database/organisms/ uniqueidentifiers/
Centre for Environmental Risk Assessment (CERA)**'s database of safety information** (formerly hosted by AGBIOS) includes not only plants produced using recombinant DNA technologies, but also plants with novel traits that may have been produced using more traditional methods, such as accelerated mutagenesis or plant breeding.	http://www.cera-gmc.org/?action =gm_crop_database
OECD Biosafety including **BioTrack Product Database**.	http://www2.oecd.org/biotech/
International Service for the Acquisition of Agri-biotech Applications, publishes Global Status of Commercialized Biotech Crops annually.	http://www.isaaa.org/
United States Regulatory Agencies Unified Biotechnology Website.	http://usbiotechreg.nbii.gov/
Information Systems for Biotechnology (ISB) provide information resources to support the environmentally responsible use of agricultural biotechnology products. They include documents and searchable databases pertaining to the development, testing, and regulatory review of genetically engineered plants, animals, and micro-organisms within the United States and abroad (supported by USDA).	http://www.nbiap.vt.edu/
Biotechnology Industry Organization	http://www.bio.org/
European Union (EU) GM food and feed, legislation, authorization, labeling, notification of existing products.	http://ec.europa.eu/food/food/ biotechnology/ gmfood/index_en.htm
European Food Safety Authority (EFSA), GMO panel.	http://www.efsa.europa.eu/en/panels/ gmo.htm
European Union Reference Laboratory for GM Food and Feed (EU-RL GMFF).	http://gmo-crl.jrc.ec.europa.eu/
GMO register: deliberate release and placing GMOs on the EU market.	http://gmoinfo.jrc.ec.europa.eu/
GMO compass is the work of independent science journalists, including the status of GMO applications/ approvals in the European Union, and other news from plant biotechnology, presented to the public in a way that is easy to understand and readily accessible.	http://www.gmo-compass.org/eng/ gmo/db/

(continued)

Table 1 (continued)

Short description	Website
Codex Alimentarius Codex Committee on Food Labeling (CCFL) Codex Committee on Methods of Analysis and Sampling (CCMAS).	http://www.codexalimentarius.net/web/index_en.jsp
The European Association for Bioindustries (EuropaBio).	http://www.europabio.org/
Proficiency Tests	
United States Department of Agriculture (USDA), Grain Inspection, Packers and Stockyards Administration (GIPSA) Proficiency Program.	http://www.gipsa.usda.gov/GIPSA/webapp?area=home&subject=grpi&topic=iws-prof-rep
The Food and Environment Research Agency providing Genetically Modified Material Analysis Scheme proficiency tests (GeMMA).	http://www.fapas.com/proficiency-testing-schemes/gemma/index.cfm
International Seed Testing Association (ISTA) Proficiency Test on GMO testing.	http://seedtest.org/en/proficiency_tests_content---1--1157.html
Bureau Inter Professionnel d'Etude Analytique (BIPEA): proficiency testing schemes.	http://www.bipea.org/en/essais.htm
Reference Materials	
Institute for Reference Materials and Measurements (IRMM).	http://www.irmm.jrc.be/html/homepage.htm
American Oil Chemists' Society (AOCS).	https://secure.aocs.org/crm/
Nippon Gene.	http://www.nippongene.com/index/english/e_index.htm
Detection Methods Databases	
Compendium of reference methods for GMO analysis. This new reference report published by the European Commission's Joint Research Centre (JRC) lists 79 reference methods for GMO analysis that have been validated according to international standards.	http://ihcp.jrc.ec.europa.eu/our_activities/gmo/gmo_analysis/compendium-reference-methods-gmo-analysis
GMOMETHODS. Searchable database based on the "Compendium of Validated GMO Detection Methods."	http://gmo-crl.jrc.ec.europa.eu/
Crop Life International Detection Methods Database.	http://www.detection-methods.com/
Chinese GMO Detection Method Database (GMDD).	http://gmdd.shgmo.org/

Legislation on GMO Labeling and GMO Detection

GMOs are officially defined in the legislation of the European Union (EU) as "organisms in which the genetic material (DNA) has been altered in a way that does not occur naturally by mating and/or natural recombination."

In the United Nations (UN) Cartagena Protocol on Biosafety, living modified organisms (LMOs) are defined as any living organism that possesses a novel combination of genetic material obtained through the use of modern biotechnology, where modern biotechnology is defined as the application of (a) in vitro nucleic acid techniques, including recombinant deoxyribonucleic acid (DNA) and direct injection of nucleic acid into cells or organelles, or (b) fusion of cells beyond the taxonomic family, that overcome natural physiological reproductive or recombination barriers and are not techniques used in traditional breeding and selection.

In the guidelines and other texts of the Codex Alimentarius Commission, an intergovernmental body to implement the Joint FAO/WHO Food Standards Programme, different terms related to GMOs are used, for example, foods derived from modern biotechnology (where the definition of modern biotechnology is taken from the Cartagena Protocol) or recombinant DNA plants, which are defined as plants in which the genetic material has been changed through in vitro nucleic acid techniques, including rDNA and direct injection of nucleic acid into cells or organelles.

Overview of the Global Situation of GMO Labeling Legislation

The first regulations specific for labeling GM food were introduced by the European Union in the late 1990s. Since then, many other countries have adopted some type of labeling policy for GM food but there are significant differences in national labeling regulations. Nevertheless some key principles of GMO labeling regulations are shared across various countries. In particular:

1. There is a large consensus among countries on the principle that GM food labeling applies only to GMOs that have undergone appropriate assessments that deem the foods safe for human consumption. Thus, labeling of GM food is not aimed to be a replacement or substitute for food safety risk assessment and management procedures; instead, it serves as an additional, potentially complementary, regulatory approach.
2. The general objective of labeling of GM food, just as with any other food labeling policy, is to inform consumers that a food product or item contains, is, or is derived from, GM products or ingredients (or does not contain/is not/is not derived from GM products). Labeling applies to all specified products regardless of their origin, both imported and domestic.
3. There is also a consensus on the fact that any labeling approach should be consistent with general food labeling principles, notably those defined by established standards of the Codex Alimentarius Commission. In particular:

    (a) Labeling should only include truthful and no misleading or confusing claims.
    (b) Food that includes identified allergens should display such allergens.
    (c) Food products, whose physical, chemical, or functional characteristics have been significantly altered, should not display this difference in a misleading manner.

Although these general principles are the basis of labeling, the specific characteristics of national GM food labeling regulations differ widely from country to country. The main common point between various national policies on GMO labeling is an agreement to require labeling for products derived from GM crops that would not be substantially equivalent to their conventional counterparts. This would concern GM products with specific novel food properties, such as soya bean oil with a specific fatty acid-profile or nutritionally enhanced rice (e.g., golden rice).

Table 2 Labeling requirements in different countries adapted from Gruere and Rao (2007)

Country	Mandatory Vs. Voluntary labeling	Product Vs. Process labeling	Threshold level (%)
EU	Mandatory	Process	0.9
China	Mandatory	Process	0
Australia, N.Z.	Mandatory	Product	1
Japan	Mandatory	Product	5
Canada	Voluntary	Product	5
U.S.A.	Voluntary	Product	n/a

There is a wide consensus among national regulators that such products should be labeled in order to inform consumers about specific novel properties of food products introduced through genetic modification.

On the other hand, for products that are considered substantially equivalent to their conventional counterparts, which includes products derived from transgenic crops with input-related traits (i.e., in practice virtually all the GM products on the market today), there are many differences in GMO labeling approaches among countries (see Table 2; Gruere and Rao 2007).

Three criteria may be used to highlight differences between the various GM food labeling approaches:

1. Type of GMO labeling (mandatory versus voluntary labeling)
2. Scope of GMO labeling (product-based versus process-based labeling)
3. Labeling threshold level

Type of GMO Labeling (Mandatory Versus Voluntary)

The first major dichotomy separates countries with *voluntary* labeling guidelines (e.g., Argentina, Canada, United States) from those with *mandatory* labeling requirements (e.g., Australia, the European Union, Japan, Korea, Brazil, China, etc.).

Voluntary labeling regulations simply define what food can be called GM or non-GM, and let the food companies decide if they want to display such information on their products. In contrast, mandatory labeling regulations require food operators (processors, retailers, sellers, and/or caterers) to display whether the specific product/ingredient contains or is derived from GM material.

Scope of GMO Labeling (Product or Process-Based Labeling)

Among countries with mandatory GMO labeling, a second major difference concerns the scope of the GMO labeling, in particular, whether the regulation requires GMO labeling:

1. Only in the presence of GM material in the final food product (e.g., Australia, New Zealand, and Japan), an approach that may be called "product-based labeling."

2. As soon as technology of genetic modification has been used in the production process (e.g., European Union, Brazil, and China) and irrespective of the presence or absence of GM material in the final food product, an approach which may be called "process-based labeling."

In the former case, only products with detectable traces of GM materials or ingredients are required to carry a GMO label. However, in the latter case, any product derived from GM crops will have to be labeled, whether it contains any traces of GM material or not. This means that refined oils are required to be labeled even if detection techniques cannot detect significant traces of rDNA or proteins in the final product.

Labeling Threshold Level

There are also differences on the threshold level for GMO labeling exemptions (i.e., the threshold below which GMO labeling is not required). The threshold levels range from 0.9% (in the European Union) to 5% (in Japan), and China with no threshold level. There may also be differences regarding application of GMO labeling thresholds to each ingredient or only to three or five major ingredients. Definition of a labeling threshold has obvious implications regarding the need for quantitative detection methods to implement legislation.

Finally, it is to be noted that international harmonization on GM food labeling has been discussed in international organizations including the Codex Committee on Food Labeling (CCFL) under the Codex Alimentarius Commission. The CCFL has been working on finding a common position on GM food labeling since the 1990s. Almost 20 years after the first discussions on an international Codex standard on GMO labeling, the CCFL has finally adopted a compilation of Codex texts relevant to GMO labeling. This standstill was due to the lack of international consensus on every aspect of the issue: from the need for a Codex guideline, its utility and possible use, to the actual content of the guideline and its implications.

The U.N. Cartagena Protocol on Biosafety also includes some provisions on labeling under Article 18 on "handling, transport, packaging and identification." However, it is important to keep in mind that its mandate focuses on transboundary movements of LMOs (i.e., either GM seeds for planting or GM commodities for food/feed processing). The Cartagena Protocol does not address products processed from GM raw materials (e.g., oil from GM soya bean, starch from GM maize). Article 18 of the Protocol, for example, establishes requirements for documentation accompanying shipments of different categories of LMOs.

Article 18(3) also foresees that the Conference of the Parties shall consider the need for and modalities of developing standards with regard to identification, handling, packaging, and transport practices. Discussions on implementation of Article 18 of the Cartagena Protocol are now leading to more and more detailed elaborations on sampling and detection of LMOs (see decision BS-V/9 taken at the MOP-5 meeting of October 2010: http://bch.cbd.int/protocol/decisions/decision.shtml?decisionid=12322).

The European Union Legislation on GMO Labeling and Traceability

The application of GMO technology is strictly regulated in the European Union and an extensive EU legislative framework on GMOs has been developed since the early 1990s. From the beginning, the key objective of the EU legislation on GMOs has been to protect health and the environment: a GMO or a food or feed product derived from a GMO can only be put on the market in the European Union after it has been authorized through a detailed EU approval process, based on a thorough scientific assessment of the risks to health and the environment.

Another clear objective of the EU legislation is to provide information to the consumers through mandatory GMO labeling for food and feed produced from GMOs.

The first major piece of EU legislation specific to GMO labeling was adopted in 1998. Regulation (EC) No. 1139/98 stated in particular that the words, "produced from genetically modified soya bean" or "produced from genetically modified maize," should appear in the list of ingredients. However, foodstuffs in which neither protein nor DNA resulting from genetic modification is present should not be subject to these additional specific GMO labeling requirements. Regulation (EC) 1139/98 introduced mandatory GMO labeling requirements for food produced from GMOs. But these additional GMO labeling requirements were limited to foods in which either protein or DNA resulting from genetic modification could be detected. These labeling requirements based on detection of DNA or protein led to the initiation of many activities related to GMO detection in the European Union.

Another major EU regulatory milestone of GMO labeling was reached in 2000 with Regulation (EC) No. 49/2000 amending Regulation (EC) 1139/98 to introduce the concepts of "adventitious presence" and of a "1% tolerance threshold" for GMO labeling exemption. The EU mandatory GMO labeling requirements were still based on detection of DNA or protein resulting from genetic modification but the notion of quantitative tolerance threshold (of 1% at that time) was introduced. These new GMO labeling requirements including a threshold level led to activities on GMO detection more focused on quantitative GMO detection.

The entire corpus of European GMO legislation was amended between 2000 and 2003, leading to the creation of an updated EU legal framework on GMOs as of 2003. In this framework two major pieces of legislation are Regulation (EC) No. 1829/2003 on genetically modified food and feed and Regulation (EC) No. 1830/2003 on traceability and labeling of genetically modified organisms and the traceability of food and feed products produced from genetically modified organisms.

Regulation (EC) No 1829/2003 on genetically modified food and feed regulates the placing on the EU market of the following products: food containing, consisting of, or produced from GMOs (GM food) and feed containing, consisting of, or produced from GMOs (GM feed).

It provides the general framework for regulating GM food and feed in the European Union and lays down in particular the EU procedures for the authorization of GM food and feed. In short a GMO food/feed cannot be placed on the EU market unless it is covered by an authorization granted according to Regulation (EC) No. 1829/2003. This authorization is based on a single-risk assessment process under the responsibility of the European Food Safety Authority (EFSA) and on a single-risk management process involving the EC and the member states through the regulatory committee procedure.

GMO Labeling

One of the key objectives of Regulation (EC) No. 1829/2003, as laid down in its Article 1, is also related to consumer information through mandatory labeling of GM food and feed, allowing consumers to make an informed choice.

It is important to note three new features of the EU GMO labeling requirements as of 2003.

1. Mandatory GMO labeling for all GM food/feed is now irrespective of the detectability of DNA or protein resulting from the genetic modification in the final product (process-based labeling). The new GMO labeling requirement therefore also includes highly refined products, such as oil obtained from GM soya bean or maize.
2. The same GMO labeling rules apply to animal feed, including any compound feed that contains or is produced from GM soya bean or maize, for instance, so as to provide livestock farmers with accurate information on the composition and properties of feed.
3. A labeling threshold of 0.9% to exempt from GMO labeling the adventitious or technically unavoidable presence of GM material in food or feed is now valid.

GMO Traceability

Products that consist of GMOs or contain GMOs and food products derived from GMOs, which have been authorized under Directive 2001/18/EC (Part C) or under Regulation (EC) No. 1829/2003, are subject to traceability requirements in the application of Regulation (EC) No. 1830/2003. Traceability is defined as "the ability to trace GMOs and products produced from GMOs at all stages of their placing on the market" (see Article 3).

Mandatory traceability of GMOs as provided for by Regulation (EC) No. 1830/2003 facilitates:

1. Control and verification of labeling claims
2. Targeted monitoring of potential effects on health and the environment, where appropriate
3. Withdrawal of products that contain or consist of GMOs where an unforeseen risk to human health or the environment is established

The traceability requirement varies depending on whether the product consists of or contains GMOs (Article 4) or has been produced from GMOs (Article 5), that is, processed products:

1. In the case of a product consisting of or containing GMOs, operators must ensure that the following information is transmitted in writing.

   (a) An indication that the product contains or consists of GMOs
   (b) The unique identifier(s) assigned to those GMOs

2. In the case of products produced from GMOs, operators must ensure that the following information is transmitted in writing to the operator receiving the product.

   (a) An indication of each of the food ingredients produced from GMOs
   (b) An indication of each of the feed materials produced from GMOs

In both cases (products consisting of GMOs or products produced from GMOs), operators must hold the information for a period of 5 years from each transaction and be able to identify the operator by whom and to whom the products have been made available. In order to respect these traceability requirements, it is important that each operator have a system in place to allow the information to be kept and to make it available to the public authorities on demand.

Exemption from the Traceability and Labeling Requirements

Conventional products – those produced without genetic modification – may unintentionally contain traces of GMOs, for example, due to cross-pollination during cultivation or due to adventitious or technically unavoidable mixing of GMO and non-GMO during harvesting, storage, transport, or processing. Taking this into account, the EU legislation has laid down a 0.9% threshold to exempt from GMO traceability and labeling requirements conventional products containing traces of GMOs below 0.9%.

More precisely Articles 12 and 24 of Regulation (EC) No. 1829/2003 (and Articles 4 and 5 of Regulation (EC) No. 1830/2003) provide that GMO traceability and labeling requirements do not apply to food and feed containing GM material in a proportion no higher than 0.9% of the food/feed ingredients considered individually, provided that this presence is adventitious or technically unavoidable (Fig. 2). Thus if the product contains different ingredients (e.g., maize, soya bean, etc.) each of them can contain up to 0.9% of GMO. More precisely, the product can contain less than 0.9% of GM maize and less than 0.9% of GM soya bean and in such a product labeling is not needed. Even if the sum of both GM ingredients exceeds 0.9% (e.g., 0.6% MON-04032-6 soya bean and 0.7% MON-00810-6 maize) labeling requirements do not apply. If there are two different GM maize present in the product, their content is added together, so if there is 0.6% of MON-00810-6 maize and 0.7% of MON-00603-6 maize, the total is 1.3% and the product must be labeled.

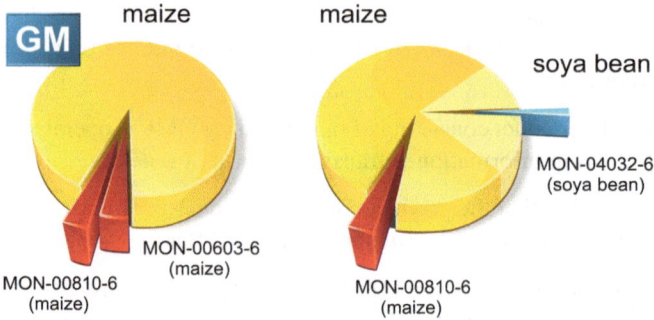

Fig. 2 Labeling requirements in the European Union are based on GMO presence in individual ingredients. Products containing 0.6% MON-00810-6 maize and 0.7% of MON-00603-6 maize have to be labeled (*left*). Products containing 0.7% of MON-00810-6 maize and 0.6% MON-04032-6 soya bean are not labeled (*right*)

In order to prove that the presence of this material is adventitious or technically unavoidable, operators must be able to supply evidence to satisfy the competent authorities that they have taken all appropriate steps to avoid the presence of such material.

In 2011, Regulation (EU) No.619/2011, the so-called low level presence regulation (LLP regulation), was adopted (European Commission 2011). This regulation applies to the detection in feed of GM material authorized for commercialization in a non-European Union countries and for which an authorization procedure has been pending for more than 3 months under Regulation (EC) No. 1829/2003 where the event-specific quantitative methods of analysis submitted by the applicant have been validated by the EU-RL GMFF and provided that the certified reference material is available. The scope of this regulation also covers GM material the authorization of which has expired. This LLP regulation introduces in particular the concept of minimum required performance limit (MRPL), as the lowest level of GM material that is considered by the EU-RL GMFF for the validation of quantitative methods. This level corresponds to 0.1% related to the mass fraction of GM material in feed and is the lowest level where satisfactorily reproducible results can be obtained from separate official laboratories when appropriate sampling protocols and methods of analysis for measuring feed samples are applied. The feed sample is noncompliant if the presence of the above-mentioned GM material is equal to or above the MRPL, measurement uncertainty being taken into account. Technical guidance from the EU-RL GMFF is available on their web page.

EU Legislation on GMO Detection

The EU legislation on GMOs includes detailed regulatory provisions on GMO detection, and in particular, defines the mandate and activities of the EU-RL GMFF in two key regulations: Regulation (EC) No. 1829/2003 on genetically modified food and feed and Regulation (EC) No. 882/2004 on official controls performed to ensure the verification of compliance with feed and food laws, and animal health and animal welfare rules.

Regulation (EC) No. 1829/2003

Regulation (EC) No. 1829/2003 confirms that submission and validation of GMO detection methods are an integral part of the EU regulatory approval process for GMOs. It provides in particular that the application for GM authorization should include, among others:

1. Methods for sampling, detection, and identification of the transformation event.
2. Samples of the food and their control samples, and information as to the place where the reference material can be accessed. Control samples mean the GMO or its genetic material (positive sample) and the parental organism or its genetic material that has been used for the genetic modification (negative sample).

Article 32 and the Annex of Regulation (EC) No. 1829/2003 lay down in particular the duties of the EU-RL GMFF. It also stipulates that applicants for authorization of GM food/feed should contribute to supporting the costs of the tasks of the Community Reference Laboratory (now EU-RL GMFF) and the ENGL.

The Annex of Regulation (EC) No. 1829/2003 in particular provides that:

1. The EU-RL GMFF is the Commission's JRC.
2. The EU-RL GMFF is assisted by the National Reference Laboratories (NRLs) referred to in Article 32, which are consequently considered as members of the consortium referred to as the ENGL.
3. The EU-RL GMFF is responsible, in particular, for:

   (a) The reception, preparation, storage, maintenance, and distribution to the members of the ENGL and to NRLs of the appropriate positive and negative control samples.
   (b) The evaluation of the data provided by the applicant for authorization for placing the food or feed on the market, for the purpose of testing and validation of the method for sampling and detection.
   (c) The testing and validation of the method for detection, including sampling and identification of the transformation event.
   (d) The submission of full evaluation reports to the EFSA.

The EU-RL GMFF also plays a role in the settlement of disputes concerning the results of the tasks outlined in the Annex.

Two other EU regulations, Regulation (EC) No. 641/2004 and Regulation (EC) No. 1981/2006, provide further detailed rules on implementation of Regulation (EC) No. 1829/2003 and the activities of the EU-RL GMFF.

Regulation (EC) No. 641/2004

Regulation (EC) No. 641/2004 provides further details on the applications for authorization of GM food and feed, including the method(s) of detection, sampling, and event-specific identification of the transformation event. In particular Annex I of Regulation (EC) No. 641/2004 on "method validation" provides detailed technical provisions on the type of information on detection methods that shall be provided by the applicant and that is needed to verify the preconditions for the fitness of the method. This includes information about the method as such and about the method testing carried out by the applicant. Annex I of Regulation (EC) No. 641/2004 also confirms that the validation process will be carried out by the EU-RL GMFF according to internationally accepted technical provisions and that all guidance documents produced by the EU-RL GMFF are to be made available.

Regulation (EC) No. 1981/2006

Regulation (EC) No. 1981/2006 provides further detailed rules specific for the EU-RL GMFF, in particular about:

1. The contribution to the costs of the tasks of the EU-RL GMFF and of the NRLs
2. The establishment of NRLs assisting the EU-RL GMFF for testing and validating the methods of detection and identification

Annex I of Regulation (EC) No. 1981/2006 lays down the minimum requirements to be fulfilled by the NRLs assisting the EU-RL GMFF (including the requirement to be accredited, or in the process of accreditation, according to EN ISO/IEC 17025). Annex II of Regulation (EC) No. 1981/2006 lists the laboratories appointed as NRLs under Regulation (EC) No. 1829/2003 to assist the EU-RL GMFF for testing and validating detection methods.

Regulation (EC) No. 882/2004

In addition to Regulation (EC) No. 1829/2003, a second key piece of EU legislation defining the mandate and activities of the EU-RL GMFF is Regulation (EC) No. 882/2004 (on official controls performed to ensure the verification of compliance with feed and food laws, animal health, and animal welfare rules).

Annex VII of Regulation (EC) No. 882/2004 lists the various European Union Reference Laboratories for food and feed and provides in particular that EU-RL GMFF is the same laboratory as referred to in the Annex of Regulation (EC) No. 1829/2003 on GM food/feed, that is, the Commission's JRC. Pursuant to Article 32 of

Regulation (EC) No. 882/2004, all European Union Reference Laboratories referred to in Annex VII (including therefore the EU-RL GMFF) are responsible for:

1. Providing NRLs with details of analytical methods, including reference methods
2. Coordinating application by the NRLs of the methods referred to in (1), in particular by organizing comparative testing and by ensuring an appropriate follow-up of such comparative testing in accordance with internationally accepted protocols
3. Coordinating, within their area of competence, practical arrangements needed to apply new analytical methods and informing NRLs of advances in this field
4. Conducting initial and further training courses for the benefit of staff from NRLs and of experts from developing countries
5. Providing scientific and technical assistance to the Commission, especially in cases where member states contest the results of analyses
6. Collaborating with laboratories responsible for analyzing feed and food in third-world countries

Pursuant to Article 33 of Regulation (EC) No. 882/2004, the member states should designate one or more NRL for each EU-RL (including therefore for the EU-RL GMFF). A list of the various NRLs responsible for GMO controls is available at http://ec.europa.eu/food/food/biotechnology/gmo_reference_lab_en.htm

Pursuant to Article 33 of Regulation (EC) No. 882/2004, the competent authority shall designate laboratories that may carry out the analysis of samples taken during official controls (official laboratories). However, competent authorities may only designate laboratories that operate and are assessed and accredited in accordance with the following European standards EN ISO/IEC 17025 on "General requirements for the competence of testing and calibration laboratories." The accreditation and assessment of testing laboratories may relate to individual tests or groups of tests.

An overview of the key regulatory texts on GMO detection with key provisions is compiled in Table 3.

Table 3 Key EU regulatory texts on GMO detection requirements

Number	Topic	Publication	Key provisions
Regulation (EC) No. 1829/2003	Genetically Modified Food and Feed	OJ L 268 18.10.2003	Community procedure for authorization of both GM food and GM feed (including one door-one key authorization process, allowing approval of a GMO under Regulation (EC) No. 1829/2003 both for food/feed uses and for cultivation).
			Mandatory labeling for all GM food and feed, irrespective of detectability of DNA or protein resulting from the genetic modification.
			0.9% labeling threshold for the adventitious or technically unavoidable presence of GM material in food or feed.

(continued)

Table 3 (continued)

Number	Topic	Publication	Key provisions
			Mandatory submission of detection methods and samples of GM food/feed, including validation by the Community Reference Laboratory (now EU-RL GMFF).
Regulation (EC) No. 1830/2003	Traceability and Labeling of GMOs and food feed produced from GMOs	OJ L 268 18.10.2003	Operators must transmit the following information to the operator receiving the product: – An indication that the product contains GMOs – The unique identifier(s) assigned to those GMOs.
Regulation (EC) No. 641/2004	Detailed rules for implementation of Regulation (EC) No. 1829/2003 on GM food feed	OJ L 102 07.04.2004	Details regarding the contents of an application for GM food feed authorization, in particular regarding method validation and reference material.
Regulation (EC) No. 882/2004	Official controls performed to ensure compliance with feed and food law	OJ L165 30.04.2004 (corri-gendum in OJ L 191 28.05.2004)	Community harmonized framework on official controls performed to ensure compliance with feed and food law. Designation and activities of Community Reference Laboratories and National Reference Laboratories (incl. on GMOs).
Regulation (EC) No. 1981/2006	Detailed rules for implementation of article 32 of Regulation (EC) No. 1829/2003 on the CRL for GMOs	OJ L 368 23.12.2006	Detailed rules concerning: - The contribution to the costs of the tasks of the Community Reference Laboratory (now EU-RL GMFF) and of the National Reference Laboratories - The establishment of National Reference Laboratories assisting the Community Reference Laboratory for GMOs (now EU-RL GMFF).
Regulation (EU) No. 619/2011	Low Level Presence of GM material in feed	OJ L 166 25.06.2011	Methods for sampling and analysis for the official control of feed as regards to presence of genetically modified material for which an authorisation procedure is pending or the authorisation of which has expired.

Organization of the Laboratory and Quality Management System

Introduction

In 1997 the Codex Alimentarius recognized the importance of harmonizing quality assurance of food-testing laboratories, setting the requirements for competence of testing laboratories involved in the import and export control of foods, in compliance with the general criteria for testing laboratories laid down in EN ISO/IEC 17025:1999 (revised as the currently EN ISO/IEC 17025:2005; Codex Alimentarius Commission 1997; International Organization for Standardization 2005d). In the European Union, laboratories can be designated by competent authorities to carry out the analysis of samples taken during official controls only if they operate, are assessed, and accredited in accordance with the EN ISO/IEC 17025 standards (European Commission 2004c).

EN ISO/IEC 17025 sets down general requirements for the competence of testing and calibration laboratories (International Organization for Standardization 2005d). The main technical requirements for detection of GMOs are detailed in the following standards on Foodstuffs – Methods of Analysis for Detection of GMOs and Derived Products:

EN ISO/IEC 24276:2006 – General requirements and definitions (International Organization for Standardization 2006)

EN ISO/IEC 21571:2005 – Nucleic acid extraction (International Organization for Standardization 2005c)

EN ISO/IEC 21569:2005 – Qualitative nucleic acid-based methods (International Organization for Standardization 2005a)

EN ISO/IEC 21570:2005 – Quantitative nucleic acid-based methods (International Organization for Standardization 2005b)

The user manual on the analysis of food samples for the presence of GMOs prepared by the IHCP contains many useful practical instructions (Joint Research Centre – Institute for Health and Consumer Protection 2006). It is available at their website in many languages (http://mbg.jrc.ec.europa.eu/capacitybuilding/documentation.htm). Additionally, recommendations for implementation of requirements of EN ISO/IEC 17025 in GMO testing were given by Žel and coworkers (Žel et al. 2006).

The EN ISO/IEC 17025 standard has two main parts: one on management requirements and another on technical requirements. It is important to note that in GMO testing, management requirements should also be considered carefully. For example, all activities during GMO testing from the initial contact with the customer to the final sample test report should be documented in a traceable manner, also ensuring protection of the data and confidentiality. All data about the sample itself and the experimental analyses can be stored as electronic documents in database(s). Appropriate backups should be considered. In this section the focus is, however, given to the key technical requirements related to detection of GMOs, especially for PCR-based methods. The section is organized in a similar manner to the standard EN ISO/IEC 17025 and the same section headings are used for each topic discussed (International Organization for Standardization 2005d). For additional technical information regarding experimental and data analysis, see the sections Homogenization and DNA Extraction, Real-Time PCR, Verification of Methods, and Measurement Uncertainty.

Testing of GMOs is composed of different steps, starting from the contact with the customer who needs complete information on the testing options, their characteristics, and constraints, to the final test report that should again be provided in a form that is understandable to customers.

Personnel

Personnel are the heart of every laboratory. They have to be reliable, precise, and motivated. Detection of GMOs is evolving constantly, therefore continuous education and training of personnel is of high importance. Laboratory management shall ensure the competence of all who operate specific equipment, perform tests, evaluate results, and sign test reports and calibration certificates. Managerial and technical tasks and their relevant responsible persons should be clearly defined, including substitutes in the case of absence of responsible personnel.

Accommodation and Environmental Conditions

One of the most important precautions in GMO testing is to prevent cross-contamination of the samples and organize the unidirectional route of the sample (Fig. 3). Therefore separate rooms (or chambers) for each of the following testing steps should be assured wherever possible: reception of the samples, preparation of samples including homogenization, extraction of DNA, PCR mix preparation, addition of the extracted DNA to the PCR reaction wells, and analysis of PCR products.

Each step has its own danger of cross-contamination. The thin dust produced during sample homogenization, or traces of previously extracted DNA are perhaps the most problematic contaminants of PCR reactions. Additional precaution should be taken when handling plasmids, especially if the laboratory prepares them itself from bacterial cultures.

Usual good laboratory practices such as changing gloves and laboratory coats, using disposable plasticware, separate reaction reagents and pipette sets, and so on for each room, significantly decrease the chance of contamination between different stages of the detection procedure. Additional measures can be adopted for further reduction of possible contamination including the use of pipettes with filters and tube opener, cleaning with nucleic acid cleansing solutions for removing DNA, and radiation with UV light before the work especially in the chamber for PCR mix preparation and the chamber where DNA is added into the reaction wells.

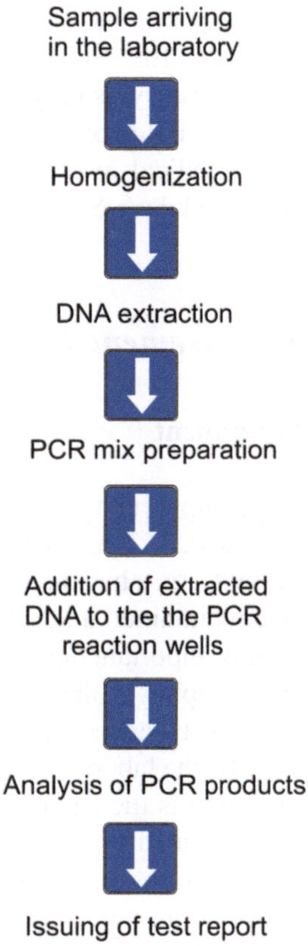

Fig. 3 Scheme showing unidirectional route of sample in GMO testing from the first contact with the customer, through analyses in the laboratory, and issue of the final test report to the customer. Wherever possible, separate rooms (or chambers) should be assured for performing each stage of the procedure

Sample arriving
in the laboratory

⬇

Homogenization

⬇

DNA extraction

⬇

PCR mix preparation

⬇

Addition of extracted
DNA to the the PCR
reaction wells

⬇

Analysis of PCR products

⬇

Issuing of test report

Another important factor to consider in the organization of a GMO testing laboratory is the temperature. Room temperature control (e.g., $23 \pm 3\,°C$) guarantees that pipetting of small volumes is not influenced by the environment.

Test Methods and Method Validation

According to EN ISO/IEC 17025, the laboratory needs a clear policy on the implementation of new detection methods. Several parameters can influence this plan, among which the most important are availability of the methods and customers' demands. Most EU laboratories implement screening methods to cover a wider range of GMOs. Quantitative event-specific methods are usually implemented to identify GMOs approved in the European Union, although some laboratories may also decide to implement methods for the detection of those GMOs on the EU market that are either unauthorized or more liable to appear. See also the section on Methods. Part of this EN ISO/IEC 17025 section is also a validation of methods, described in detail in the section, Verification of Methods.

Estimation of measurement uncertainty (MU) is also a requirement covered under this EN ISO/IEC 17025 section. The establishment of the system for MU in the laboratory usually requires some considerable effort, especially for understanding the whole concept of MU. Once the system is in place, the effort remains only for periodical re-evaluation of the estimated MU values taking into account new analytical data. See the section, Measurement Uncertainty.

Measurement Traceability and Reference Materials

Equipment

All equipment to be used for routine GMO detection should be calibrated and validated. If the laboratory uses equipment supplied by another organization, for example, as a backup of their own equipment, it has to ensure that this equipment is also maintained under ISO/IEC 17025 requirements.

It is important that all equipment meets the laboratory's specification requirements and complies with the relevant standards. Calibration of equipment (PCR apparatus, pipettes, balances, etc.) can be performed by a qualified service provider or can be done by the laboratory itself, but the quality of the calibrations must be assured within the lab. It is the laboratory's responsibility to decide which calibrations can and will be performed by its own personnel and which by an external service; sometimes the best decisions are made through discussions with the service provider. An intermediate solution, and in our view very practical, is that calibrations are performed by a qualified service provider at regular, somewhat longer but still appropriate, intervals, and that between these, intermediate checks are done by the laboratory's personnel.

Reference Materials

General

Reference materials are used as positive controls for qualitative and quantitative purposes as described in the Real-Time PCR section. A certified reference material (CRM) for which sufficient information on its quality and origin is available is preferred to a reference material without a certificate or where the certificate is lacking essential information (Žel et al. 2008). More detailed technical provisions for development and production and further requirements for reference materials, such as homogeneity, stability, storage, and certificate information are described in Annex II of Regulation (EC) 641/2004 (European Commission 2004a).

Reference material for DNA-based methods is a material containing the analyte. This can be a powdered material (e.g., flour from seeds) containing the analyte, DNA extracted from material containing the analyte, or a plasmid containing the specific analyte nucleotide sequence (Fig. 4). Preferably CRMs certified for the

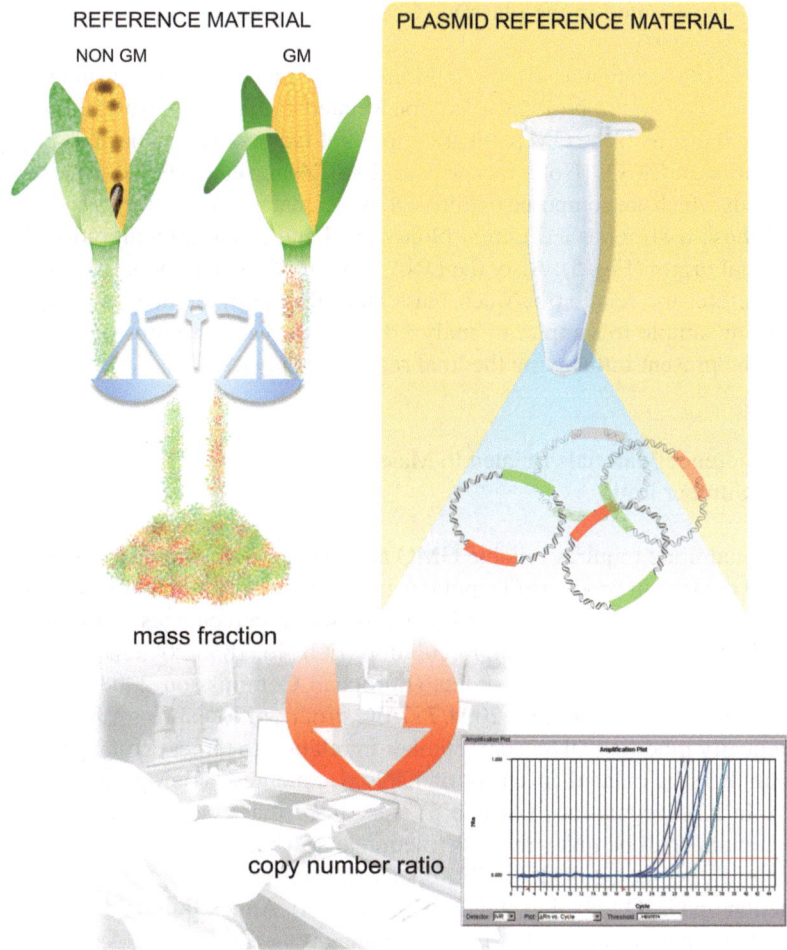

Fig. 4 Different kinds of reference materials are available, such as powdered reference material (*left*) or plasmids (*right*). Powdered reference materials are usually certified for different mass fractions (e.g., 1%, 5% MON-00810-6) prepared by quantitative mixing of non-GM powder and GM powder. After additional copy number studies they can be certified also for copy number ratio between event-specific and taxon-specific sequences, as is automatically calculated for dual plasmid reference materials with a 1:1 ratio between targets

presence of the analyte should be used, but if they are not available, a positive control sample (e.g., from proficiency testing schemes) can be used as reference material. It is important to be aware that CRMs are certified for the presence of a given event and not for the absence of other events. Trace contaminations of CRMs by other GM events were detected on a regular basis in our lab.

Before using plasmids as reference material, one must carefully ensure that the plasmid or the amplicon DNA sequence incorporated in the plasmid will be fit for the required purpose, namely that the method which will be used is targeting the sequence incorporated in the plasmid (Codex Committee on Methods of Analysis and Sampling 2010).

One special topic that needs to be addressed in relation to reference materials and samples tested is that of biological factors. Related to plants, zygosity, tissue ploidy, and parental origin of the GM plant are important factors that can have an impact on quantification of GMOs (Holst-Jensen et al. 2006; Zhang et al. 2008a). An example is that of seeds which are composed of different tissues, endosperm, embryo, and pericarp. Each of these tissues has a different ploidy level and has a different ratio of maternal/paternal origin (Fig. 5). Also, the DNA content of each of these tissues is different. Therefore, the relation between mass and DNA copy number is complex and varies from sample to sample. In analyzed samples, varying ratios of different tissues can be present influencing the final result of analyses.

Using Reference Materials Related to Mass, Mass Fraction, or Copy Number Ratio

International trade requires reliable GMO analysis for comparable measurement of the GMO content in products (Trapmann et al. 2010). At international and national levels the presence of GMOs should be expressed in percentages, but units are not specified. In the European Union, Regulation (EC) No. 1830/2003 has not specified the measurement units for thresholds (European Commission 2003b). In 2004 a Commission Recommendation 2004/787/EC proposed to express "the percentage of genetically modified DNA copy number in relation to target taxon-specific DNA copy numbers calculated in terms of haploid genomes" (European Commission 2004b). In LLP regulations (Article 3) it is stated that "the certified value of the GMO content shall be given in mass fraction and, where available, in copy number per haploid genome equivalent" (European Commission 2011; see the section on Legislation on GMO Labeling and GMO Detection). Annex II of the LLP regulation further states that "when results are primarily expressed as GM-DNA copy numbers in relation to taxon-specific DNA copy numbers calculated in terms of haploid genomes, they shall be translated into mass fraction." The lack of coherence between legal requirement and approaches for detection of GMOs was well described a few years ago (Holst-Jensen et al. 2006).

Expressing the GMO quantification result in GMO mass fraction or GMO copy number ratio requires a different type of calibrant. It is recommended that CRMs prepared on mass/mass scale and certified for the ratio between GMO and non-GMO counterparts of the same species be used for testing with the expression of the result in mass fraction. More information about both types of reference materials and their use for quality control and calibration can be found in different publications from the Institute of Reference Materials and Methods (IRMM), JRC, and EC (Corbisier 2007; Trapmann 2006; Trapmann et al. 2010).

Laboratories using PCR and qPCR can only measure DNA target copy numbers. As described later (see the section, Real-Time PCR), copy numbers are measured in CRM dilution series using qPCR, and these values are used to build a calibration

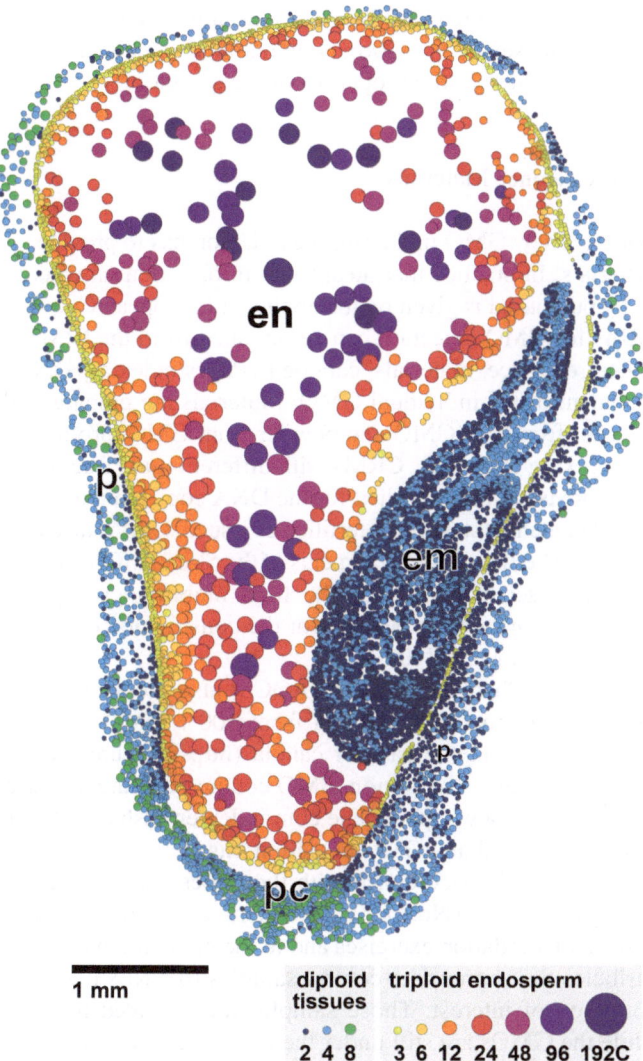

Fig. 5 Different maize seed tissues have different ploidy levels. The GM-DNA copy number of the product depends on the presence of different tissues. Graph of spatial distribution of endoreduplication in maize caryopsis. In situ DNA content of nuclei is shown in the median longitudinal tissue section of maize caryopsis 16 days after pollination (W22 inbred line). Nuclei of different endopolyploidy classes have been color-coded and the diameter of symbols is linearly related to the diameter of measured nuclei. *em* embryo, *en* endosperm, *p* pericarp, *pc* placento-chalazal layer

curve for determining the copy numbers in routine samples. Although sufficient information about CRMs is sometimes available for transformation of mass/mass ratio to relative copy number (e.g., zygosity, tissue ploidy, parental origin of the GMO, extractability of DNA from material, etc.), there is a lack of information on the composition of those samples tested. For these samples, there is obviously no information on the above-mentioned factors that influence GMO copy number. As proposed by Holst-Jensen et al. (2006), copy number-based quantification could be most consistent regardless of the properties of the sample.

Availability of Reference Materials

The applicant for a new GMO in the European Union has to provide accessibility to reference materials. In practice that means that information on where the reference material can be purchased is given by the applicant (European Commission 2003a).

Many CRMs for GMO detection can be obtained from the IRMM (http://irmm. jrc.ec.europa.eu/reference_materials_catalogue/Pages/index.aspx). CRMs are certified for the content of the individual GMO. Materials are prepared by quantitative mixing of powder from non-GMO and powder from GMO, produced from ground seeds by a dry-mixing technique. CRMs with different mass fractions are available. Some of these CRMs are also certified for the DNA copy number ratio. Uncertainty of CRMs' GMO content is stated and certificates issued. IRMM also offers a plasmid, certified for the ratio between DNA fragments of the 5'-end MON-00810-6 transgene–host plant junction sequence and of the high mobility group gene A (*hmgA*) (ERM-AD413). It is intended to be used for the construction of a calibration curve for the quantification of MON810 maize.

The American Oil Chemists' Society (AOCS) also produces certified reference materials, available as powder (prepared from 100% GMO seeds) or leaf tissue DNA and also as seeds in the case of canola (https://secure.aocs.org/crm/index. cfm). Some reference materials of non-GMO counterparts are also available.

Nippon gene (http://www.nippongene.com/index/english/e_index.htm) produces plasmid reference material and GMO detection kits.

The EU-RL GMFF, in accordance with the duties and tasks established in the annex to the Regulation (EC) No. 1829/2003 (European Commission 2003a), further to the conclusion of validation exercises and to the publication of the corresponding reports, distributes the appropriate control samples to NRLs that have replied to a call for expression of interest. Those samples are intended to allow for control purposes while the GMOs are still under the authorization process and no certified reference material is yet available (Žel et al. 2008).

Handling of Reference Materials

Handling of reference materials in the laboratory should be defined; including the description of the limitation of use (e.g., some reference material may only be suitable for qualitative analyses). Reference material for qualitative analyses can be a sample with known presence of GMO, where the target presence is confirmed by certificate (CRMs; preferably), by interlaboratory comparison or intralaboratory trials.

Rarely, it can happen that for some reason (e.g., problems occurring during transportation or storage) a reference material does not give expected results (GMO presence and content); therefore it is recommended to take precautions when a reference material is used for the first time. In quantitative analyses, new reference materials can be tested through the use of quantification control samples (see below). To avoid degradation of reference materials during their storage, they should be kept at below −15 °C and repeated freezing/thawing should be avoided.

According to our experience, integrity of nonopen reference materials is not affected for at least two years, thus expiration time can be extended to this period, unless the certificate of reference material specifically states a risk of degradation in a shorter period. From our experience, DNA extracted from reference material has two years' expiration time if kept at below −15 °C.

Sampling

Sampling is beyond the scope of this book, which is meant to deal with laboratory-related issues in GMO detection. The subsampling procedure performed following homogenization of the samples is given in the section, Homogenization and DNA Extraction.

Handling of Test Items

Samples

Immediately upon arrival at the laboratory, samples are labeled with unique identification tags to assure traceability of the sample and confidentiality of the customer's data. This label accompanies the sample throughout the testing process, from its entrance into the laboratory, through experimental analyses, and to the final test report.

Systematic storage of samples and extracted DNA ensures the possibility of repeating the analyses if needed. The following system for sample and extracted DNA storage is proposed and is working well within our system, even though other systems are possible.

After the homogenization step, an aliquot of homogenized sample is stored in a 15-ml tube (app. 5 g) at 2–8 °C. Homogenized sample aliquots are stored for at least three months. Two parallel DNA extractions are performed from each sample and extracted DNA stored at below −15 °C for at least three months after extraction.

Primers and Probes

A system for treatment, storage, and purchase of chemicals has to be clearly established to enable the repeatable performance of chemicals in subsequent analyses. Primers and probes are among the key chemicals for qPCR analyses; therefore the following precautions are recommended. Each time new primers or probes are purchased they need to be tested in parallel with the ones already in use in the laboratory to assure that they work properly. Key qPCR parameters (Cq values, amplification efficiency, and shape of the amplification curve) are compared between runs using new and old chemicals, and acceptance criteria described in section Real-Time PCR have to be fulfilled. Primers and probes are kept at below −15 °C, in the dark. Expiration time for lyophilized primers is unlimited. For primers provided in solution, the expiration date proposed by the producer is six months after receipt. According to our experience, the expiration date of diluted primers and probes can be extended if a regular monitoring of chemical performance is done.

Lyophilized primers and probes should be diluted to stock concentration (e.g., 100 µM). To prevent repeated exposure of probes to the light, repeated freezing/thawing of primers and probes, and potential cross-contamination, it is preferable that a part of this stock is further diluted into working concentrations needed for PCR mix and stored in aliquots (e.g.,100 µl).

It is possible to combine the forward primer, reverse primer, and probe solution for each assay in a single solution to stock concentrations needed for the PCR mix. Extensive evaluation of such a system has been performed in our laboratory and no difference was observed between combined and individual solutions (data not published). Preparation of such combined primers/probe solutions results in reduced pipetting during qPCR setup and decreases the chance of mistakes in the preparation of a qPCR mix.

Assuring Quality of Testing

To monitor possible trends or bias in sample analyses, different control measures can be taken. Participation in proficiency tests, use of control charts, and regular use of analytical controls is appropriate and sufficient for GMO testing. Additional analysis reliability can be achieved by using a standardized qPCR plate setup.

Proficiency Tests

Participation in proficiency tests is one of the main requirements of EN ISO/IEC 17025 to show laboratory competence in analyses (International Organization for Standardization 2005d). Participation in proficiency testing schemes is crucial for the independent assessment of laboratory performance. Participation in proficiency tests depends on the methods introduced in the laboratory and the type of samples analyzed. Laboratories can use many methods for routine testing and not all of them can be assessed as each proficiency test covers only a limited number of GMOs. Therefore it is recommended to prepare a plan of participation in proficiency tests and to assess individual methods periodically. It is also important to cover different types of matrices that are subject to testing during routine analyses. Different proficiency testing schemes are available and therefore selection can be done regarding the needs of individual laboratories.

The EU legislation also makes participation in comparative testing mandatory for NRLs. Article 32 of Regulation (EC) No. 882/2004 states in particular that the EU-RLs are responsible among others for coordinating application by the NRLs of the methods, in particular by organizing comparative testing and by ensuring an appropriate follow-up of such comparative testing in accordance with internationally accepted protocols, when available. Proficiency testing programs for GMO detection are provided by different institutions, some of them listed in the following paragraphs.

The EU-RL GMFF organizes comparative testing rounds in which participation is mandatory for all NRLs nominated under Regulation (EC) No. 882/2004 and Regulation (EC) No. 1829/2003. Tests are done on events in flours. In February 2011 the EU-RL GMFF published its first "Comparative Testing Report on the Detection and Quantification of Maize."

The Food and Environment Research Agency (FERA, previously CSL) organizes the GeMMA Scheme including: maize and soya bean events in flours, baked products, and animal feed; soya bean in processed matrices; and soya bean, maize, rice, rapeseed, sugar beet, and potato events in DNA samples.

The USDA/GIPSA Proficiency program provides flour samples fortified with various combinations and concentrations of transgenic maize and soya bean events for qualitative or quantitative analysis.

In the ISTA Proficiency Test on GMO Testing, participants are required to quantify the content of GM seeds in samples either by a subsampling quantification (semi-quantitative test) or by a quantitative test. In the case of quantitative tests, results must be reported as percentages, either related to the number of seeds or to the mass of seeds or DNA copy numbers. The participation in the ISTA Proficiency Tests on GMO is mandatory for ISTA accredited member laboratories.

A laboratory should evaluate the results obtained in proficiency tests, and investigation should be made in the case of noncompliant performance. The reasons for incorrect results should be precisely determined and corrections made accordingly. It must be noted that these different proficiency test programs do not use the same evaluation system to assess the results, making it difficult to compare laboratory results between different proficiency tests. In EU-RL GMFF comparative testing rounds, z-scores are calculated on the basis of the assigned value by the test item provider and the robust mean of the participants' results. In GeMMA proficiency schemes in particular, the z-scores are based on the results obtained by the participant laboratories, whereas in USDA/GIPSA proficiency schemes z-scores are simply based only on the values assigned according to preparation of the test samples (in mass fraction). In this case it is recommended to replot the laboratory results against the results of the other laboratories in order to get a better estimate of laboratory performance. The difference in zygosity between the reference material used by the laboratory to build standard calibrating curves, and the test samples provided by the proficiency test schemes can also significantly influence laboratory performance in proficiency tests.

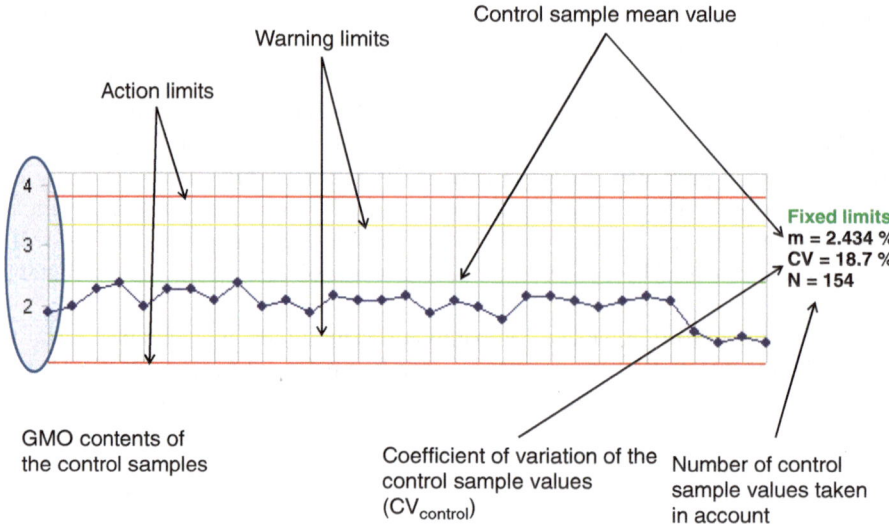

Fig. 6 Example of a control chart for quantification control sample. GMO content measured in a given quantification control sample is plotted over time in a control chart (*blue dots*). The mean control sample value (*m*) is calculated (in this example 2.434%) as well as the coefficient of variation of the control sample values in time (CV$_{control}$, in this example 18.7%). From *m* and CV$_{control}$ the warning (*yellow lines*) and action limits (*red lines*) are estimated. These limits as well as the CV$_{control}$ value are used to monitor the quality of the control sample quantification

Control Charts

Control charts are very useful tools for quality control in a laboratory. They can help to detect possible trends in analytical results such as longer-term deviations from true value. These charts can be used for monitoring equipment performance (e.g., calibration of pipettes, laboratory temperature, refrigerator and freezer temperature, etc.) and for monitoring quantification accuracy.

To control the quality of quantitative analyses, positive samples for different GM lines with known GMO contents should be regularly analyzed in parallel with quantification of samples for the customers, the quantification control system. The GMO contents of these control samples are followed over a certain period of time allowing identification of possible trends (increasing or decreasing values, punctual changes) in the results of the quantification control sample analysis. Possible deviations can be observed and related to the analysis of samples for customers.

GMO contents calculated for each quantification control sample are plotted in time in control charts, also known as Shewhart charts (e.g., in a spreadsheet program-based graph) allowing further evaluation of the results according to set limits (Fig. 6). Two types of limits are defined: the warning limit is set at the mean control sample GMO value +/−2 CV$_{control}$, and the action limit is set at the mean control sample GMO value +/−3 CV$_{control}$, where the coefficient of variation of the control sample GMO value (CV$_{control}$) is the standard deviation in time of the GMO values measured for the control sample divided by the mean control sample GMO value. $CV_{control}$ is expressed as percentage.

If a series of control sample values passes the warning limit, a closer look should be given to the results and monitoring of the quantification control sample should be performed more often to identify if a trend (over- or underestimation) exists in the results of the laboratory analyses leading to bias in quantification. The same action should be taken if the $CV_{control}$ of the control sample is higher than 33%, showing a great instability of the laboratory results when measuring the control sample GMO contents.

If a control sample value crosses the action limit, immediate investigation should be made to analyze the cause of this deviation and correct the problem.

Analytical Controls

The use of analytical controls is essential to ensure quality of the test performance and results. Different types of controls used during analyses should be clearly specified. These may include positive and negative controls, their detailed content, and the extent to which they should be used together with the interpretation of the obtained values (Codex Committee on Methods of Analysis and Sampling 2010). See more details in the sections, Homogenization and DNA Extraction and Real-Time PCR.

Reporting the Results

Reports on analyses of results have to involve all administrative and technical data described in EN ISO/IEC 17025 (International Organization for Standardization 2005d) as well as EN ISO/IEC 24276 (International Organization for Standardization 2006), EN ISO/IEC 21569 (International Organization for Standardization 2005a), and EN ISO/IEC 21570 (International Organization for Standardization 2005a). Test reports should contain, for example, information needed to identify the laboratory sample, including any particular information related to the laboratory sample, statement about the date and type of sampling procedures used, date of receipt, size of the laboratory sample and test sample, results according to the requirements of the specific method and the units used to report the results, person responsible for the analysis, and any deviations, additions to, or exclusions from, the test specifications. More details are given in the section, Real-Time PCR.

Methods

Introduction

Testing of GMOs is based on the detection of rDNA introduced during the transformation process or on the recombinant proteins expressed in the GMO. Recombinant proteins or rDNA are targets that differentiate GMOs from their non-GMO counterparts (see the Introduction).

DNA-Based and Immunological Methods

Detection of rDNA is performed using molecular, mostly PCR-based methods, whereas proteins are detected by immunological methods. There are differences between both approaches, determined by the general characteristics of the analytes themselves. Although rDNA is present in each GMO, expression of recombinant proteins can vary a lot or the protein is even not produced in the GMO (e.g., gene silencing mechanism). Event-specific DNA-based methods enable differentiation among individual GMOs, and protein-based methods identify the recombinant protein that can be produced in different GMOs. Practically, the same rDNA can be introduced into different GMOs resulting in the presence of the same protein in different GMOs, preventing their individual identification. On the other hand the same rDNA is never introduced in the same position of the plant genome. Therefore detection of junction regions between rDNA and plant DNA enables specific identification of the GMO. The advantage of the protein-based methods is their lower price in comparison with DNA-based methods, especially when strip tests are used. Protein strip tests represent a useful tool to trace proteins in raw materials such as seeds and leaves from crop plants. In food/feed products, the protein strip test applicability in GM tracing is restricted to samples containing sufficient GM material derived from plant tissues where the recombinant protein is expressed and limited by the inherent physicochemical properties of the proteins themselves (thermostability, quenching interference; Van den Bulcke et al. 2007).

In the European Union and some other countries, legislation is based on individual GMOs, and the thresholds for their content are set; therefore GMO testing laboratories use DNA-based methods that allow precise quantification of GMOs in samples. Conventional PCR can be used for qualitative analysis, whereas qPCR can be used for qualitative and quantitative purposes. The following sections focus only on qPCR methods.

The homogenization of samples and DNA extraction are described in detail in the section Homogenization and DNA Extraction, and performance of qPCR in Real-Time PCR.

Singleplex, Multiplex Methods

Singleplex DNA-based methods directed at a single target are the most generally accessible and thus most often used in routine GMO diagnostics. Recently, development of multiplex qPCR methods for simultaneous detection of more than one target was reported, but so far these methods have rarely been validated in collaborative trials (Bahrdt et al. 2010; Dorries et al. 2010; Gaudron et al. 2009; Waiblinger et al. 2008b). The duplex qPCR method targeting the DNA sequences from Cauliflower Mosaic Virus 35 S promoter (P-35S) from Cauliflower Mosaic Virus and terminator (T-NOS) from *Agrobacterium tumefaciens* is to our knowledge the only example of multiplex qPCR detection methods validated in collaborative trials

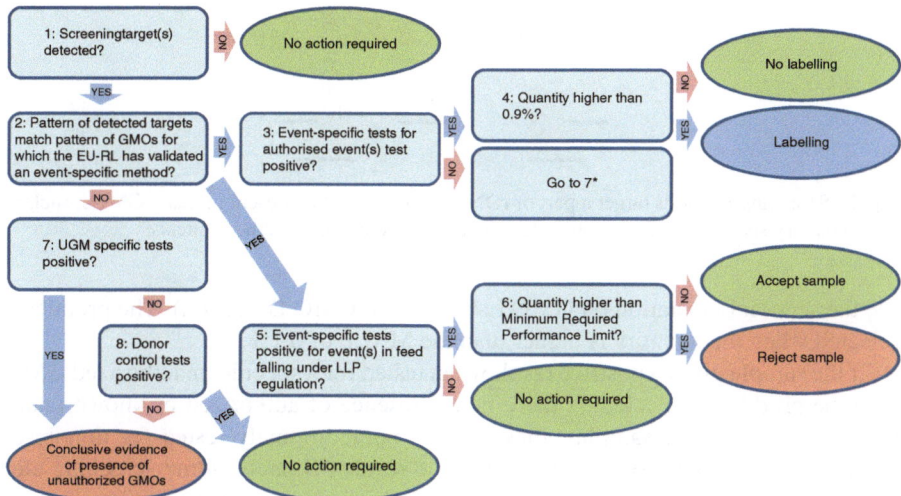

Fig. 7 Scheme of GMO testing in EU official laboratories, simplified version (see text for detailed explanation). *If sample is feed it can be under LLP regulation (no. 5)

(Waiblinger et al. 2008b). The main limitation for multiplex methods development is the existence of possible interferences during simultaneous multiplication of amplification targets. See also the section New Challenges.

Detection Procedure Using DNA-Based Methods

Scheme of GMO Testing

Testing of GMOs is usually performed in a stepwise system (Holst-Jensen 2007). Here we describe the scheme used for GMO detection in EU official control laboratories, including decision making according to obtained results (Fig. 7). It is based on the current work of the ENGL working group on unauthorized GMOs. In the first step the screening tests using screening methods are performed (see below in this section for more details). If a screening result is negative, the analysis is concluded and no action on the sample tested is required. If the result of the screening test is positive, it is not known which GMO is present, therefore a second step is needed to identify the specific GMOs in the sample. After their identification using event-specific tests, validated by the EU-RL GMFF, the legal status of the GMOs present in the sample is determined.

If all GMOs present are authorized, a third step is performed: quantification of each individual GMO detected in the sample to define whether its content is above or below the labeling threshold set in legislation (0.9%). If the content is above the threshold, the product has to be labeled. Quantification is performed also if the GMO identified in feed samples falls under LLP regulation (European Commission 2011;

Fig. 8 Screening methods target a part of rDNA sequence that is present in many GMOs, such as the regulatory sequences of promoters, terminators, or construct-specific sequences

see the section Legislation on GMO Labeling and GMO Detection). The product is rejected when the quantity is higher than the MRLP of 0.1%.

If the sample gives a positive result when tested for a specific unauthorized GMO then the product is rejected. If there is no presence of authorized or known unauthorized GMO in the samples, they have to be additionally tested for donors of screening elements to exclude false positive results. Namely, some screening elements in GMOs come from organisms, which can be present in the sample, such as the most common screening element P-35S or T-NOS. If the donor control test is positive we can conclude that the positive screening test occurred due to the presence of donor organisms in the sample and the sample has no GMO present, so no action is required. A test for donors of screening elements to exclude false positive results is also possible in previous steps.

If the donor control test is negative, there is conclusive evidence for the presence of unauthorized GMOs. However, in the absence of an event-specific test, more investigation (e.g., sequence analysis) is needed to determine the precise nature of the GMOs present.

Screening

Different parts of the rDNA sequence are targeted in individual analytical steps of GMO testing. Screening methods target a part of the rDNA sequence that is present in many GMOs, such as the regulatory sequences of promoters, terminators, and so on (Fig. 8). The goal is to cover as many GMOs as possible in order to determine their presence or absence and at the same time limit the cost of analysis. The most commonly used screening targets are the P-35S and the T-NOS. The targets of screening methods are selected based on information regarding the genetic elements composing different GMOs. Construct-specific methods that usually span over two or more genetic elements, such as a promoter and its relevant trait sequence, can also be used during the screening step, usually with narrower specificity (Fig. 8).

With numerous GMOs on the market, the selection of screening elements is becoming more complex. Therefore the matrix approach, which considers each GMO as a combination of genetic elements in a given order and utilizes a smart selection of the screening elements to be targeted, is recognized as the most appropriate approach (Kralj Novak et al. 2009; Querci et al. 2010; Van den Bulcke et al. 2010; Waiblinger et al. 2008a, 2010). Appropriate selection of screening methods

Table 4 Comparison of two screening approaches on sample with possible presence of 12 GM maize lines. Use of two screening elements (for P-35S and T-NOS) in the first step, results in testing of 12 GM maize lines in identification step, and use of five screening elements in the first step (P-35S, T-NOS, CTP2-CP4EPSP, *bar*, and 35S-pat; Waiblinger et al. 2010) results in testing of only five GM maize lines (in gray) in the second step if only P-35S and T-NOS are detected in the sample

GM maize lines	P-35S	T-NOS
59122 (DAS-59122-7)	+	–
Bt11 (SYN-BT011-1)	+	+
Bt176 (SYN-EV176-9)	+	–
GA 21 (MON-00021-9)	–	+
MIR604 (SYN-IR604-5)	–	+
MON 810 (MON-00810-6)	+	–
MON 863 (MON-00863-5)	+	+
MON 88017 (MON-88017-3)	+	+
MON 89034 (MON-89034-3)	+	+
NK 603 (MON-00603-6)	+	+
T25 (ACS-ZM003-2)	+	–
TC1507 (DAS-01507-1)	+	–

GM maize lines	P-35S	T-NOS	CTP2-CP4EPSPS	*Bar*	35S-pat
59122 (DAS-59122-7)	+	–	–	–	+
Bt11 (SYN-BT011-1)	+	+	–	–	+
Bt176 (SYN-EV176-9)	+	–	–	+	–
GA 21 (MON-00021-9)	–	+	–	–	–
MIR604 (SYN-IR604-5)	–	+	–	–	–
MON 810 (MON-00810-6)	+	–	–	–	–
MON 863 (MON-00863-5)	+	+	–	–	–
MON 88017 (MON-88017-3)	+	+	+	–	–
MON 89034 (MON-89034-3)	+	+	–	–	–
NK 603 (MON-00603-6)	+	+	+	–	–
T25 (ACS-ZM003-2)	+	–	–	–	+
TC1507 (DAS-01507-1)	+	–	–	–	+

can contribute to lower the number of tests needed in identification steps, reducing the workload and cost of analyzes in subsequent steps (Table 4).

Recently the combination of five DNA target sequences was proposed as a universal screening approach (Waiblinger et al. 2010). This combination enables screening for the presence of at least 81 authorized and unauthorized GM plant events described in publicly available databases. This practical approach combines the use of P-35S and T-NOS screening methods with the two construct-specific methods for a construct containing the 5-enolpyruvylshikimate-3-phosphate synthase gene from *Agrobacterium tumefaciens* sp. strain CP4 (CTP2-CP4EPSPS), a construct containing the phosphinothricin acetyltransferase gene (P-35S-pat), and the method targeting the *bar* gene from *S. hygroscopicus*.

In order to help with the efficient selection of screening methods, bioinformatics tools have been developed to either support design of the screening step in the analysis or to support interpretation of screening step results, reducing workload

Fig. 9 Event-specific methods allowing unique identification of the individual GMO present in the sample, usually targeting the nucleotide sequence at the junction between the plant host genome and the rDNA

and cost of analyses (Kralj Novak et al. 2009; Van den Bulcke et al. 2010; see the section, New Challenges).

It is important to mention that for a given genetic element, several specific methods were developed that target different parts of the screening element's sequence, as is the case with the P-35S target (Bonfini et al. 2007). It is crucial that during validation of screening method selectivity/specificity is confirmed on as many individual GMO events as possible (as described in Verification of Methods). Also, after implementation of the method in a laboratory, selectivity/specificity should be regularly tested for new GMOs entering the market to update knowledge regularly on coverage of GMOs by individual screening methods.

One should keep in mind that small differences in the nucleotide sequence of a given genetic element (such as single nucleotide polymorphism, SNP) can be observed in individual GMOs. Such SNPs, if located in the sequence targeted by a screening method, could lead to reduced sensitivity of the method for a particular GMO and inaccurate quantification results (Morisset et al. 2009).

Identification

Identification of GMOs is performed with event-specific methods that target the specific part of the individual GMO, allowing unique identification of the GMO present in the sample (Fig. 9). Usually event-specific methods target the nucleotide sequence at the junction between the plant host genome and the rDNA (Holst-Jensen et al. 2006).

Quantification

For the quantification of GMOs, taxon-specific methods targeting sequences confined to the particular taxon of interest are needed. Quantification is done by relating the content of a taxon-specific sequence to the content of rDNA, determined by an event-specific method, in the sample (Fig. 10). In this way, the percentage of GMOs in relation to individual plant species or taxon is evaluated in accordance to legislation currently in place.

Fig. 10 Quantification of GMO is done by measuring the ratio between taxon-specific sequence and event-specific sequence

GMOs with Stacked Genes

The so-called stack GMOs (or stacked gene events) have more than one genetic construct (rDNA) incorporated in the plant genomic DNA. They can be the result of the intended introduction of several constructs during the transformation process or of the crossing of individual GMOs (Holst-Jensen et al. 2006). With routine analyses using PCR-based methods, it is not possible to differentiate between the presence in the sample of different individual GMOs (e.g., MON-00863-5 maize, MON-00810-6 maize, MON-00603-6 maize) and their stacks (MON-00863-5 × MON-00810-6 × MON-00603-6 maize). The consequence is that analysis of a sample containing a stacked GM event may result in a higher measured percentage of GMO content due to the sum of the contents for each individual event. As an example, the analysis of a sample containing 0.8% of stacked GMO MON-00863-5 × MON-00810-6 × MON-00603-6 maize will result in the detection of the three individual GMOs (0.8% of MON-00863-5 and 0.8% MON-00810-6 and 0.8% MON-00603-6 maize) with a total GMO content of 2.4% in relation to the maize ingredient, when the actual GMO content is 0.8%.

Availability of Methods

The laboratory shall use test methods that meet the needs of the customer and are appropriate for the tests it undertakes (International Organization for Standardization 2005d). Preferably, methods published in international, regional, or national standards shall be used. The basic principles for the detection of GMOs and derived products using nucleic acid-based methods are described in four standards (International Organization for Standardization 2005a, b, d, 2006).

Some methods for GMO detection are included in these documents as annexes, but most GMO detection methods are not included in the standards.

The CCMAS guidelines set the quality criteria that a detection method should achieve for reliable analyses of food samples (Codex Committee on Methods of Analysis and Sampling 2010). The goal of these guidelines is to support "the establishment of molecular and immunological methods for detection, identification and quantification of specific DNA sequences and specific proteins in foods, which produce results with comparable reproducibility when performed at different laboratories."

In addition to the four above-mentioned ISO standards, there are different available sources listing developed GMO detection methods. For example, the Chinese GMDD is a very comprehensive method database, which is part of the Shanghai GMO platform (http://gmdd.shgmo.org/). It provides detailed information of nucleic acid-based methods and protein-based methods, including primer and probe sequences, amplicon length, endogenous reference gene primers, validation information, PCR programs and references, and so on. In addition, the database also contains information about rDNA sequences and certified reference materials (Dong et al. 2008).

Another very valuable source of validated quantitative event-specific methods is the website of the EU-RL GMFF (http://gmo-crl.jrc.ec.europa.eu). The EU-RL GMFF is responsible for testing and validating the methods submitted by applicants to detect and identify the transformation events applied for authorization in the European Union. At the end of a validation exercise, the methods for identifying individual GMOs and the validation report are published on the EU-RL GMFF website and made available for further use in control laboratories testing for GMOs.

According to the Procedure Manual of the Codex Alimentarius Commission, a reference method is the one designated method recommended for use in cases of dispute and for calibration purposes (Codex Alimentarius Commission 2010). Following this definition, the quantitative event-specific methods validated by EU-RL GMFF are gaining the status of reference methods for GMO detection. Accordingly, the EU-RL GMFF and the ENGL have recently published a compendium of reference methods for GMO analysis (European Union Reference Laboratory for GM Food and Feed and GMO Laboratories 2010). This document aims at providing a list of reference methods for GMO analysis that have been validated in a collaborative trial according to the principles and requirements of ISO 5725 and/or the IUPAC protocol.

In May 2011 the Commission's JRC also launched GMOMETHODS, an online GMO detection method database. The content of the GMOMETHODS database is based on the "Compendium of Reference Methods for GMO Analysis." Both are publicly available at http://gmo-crl.jrc.ec.europa.eu/gmomethods/.

On the EU-RL GMFF website, one can also find data and methods regarding detection of GMOs that are not authorized in the European Union, but the presence of which have been identified on the EU market (e.g., CDC Triffid flax-FP967, rice Bt 63, maize Event 32, LLRICE601, maize BT10). In addition, a document is in preparation by the ENGL working group on unauthorized GMOs and will be published as a "JRC Scientific and Technical Report from the JRC." The document will be available on the JRC website (http://jrc.ec.europa.eu/). These guidelines propose approaches for the detection of unauthorized GMOs and reporting of analytical results.

There are a large number of publications reporting the development and, sometimes, in-house validations of PCR-based GMO detection methods. However, it is important to critically evaluate their validation status before implementing them in the laboratory for routine analyses of samples. In this sense, a recent work from Kodama and collaborators can be useful: the authors have extensively studied the results of a large number of GMO method collaborative trials (Kodama et al. 2010).

Verification of Methods

Introduction

Several measures need to be implemented by a laboratory to ensure reliability and quality of the analytical output (CRL-GMFF 2008). One of these measures is the use of methods that have been subjected to interlaboratory validation following internationally recognized standards (CRL-GMFF 2008; European Commission 2004c). In the European Union, each applicant must provide a qPCR method specifically detecting the GMO to be approved for import and processing, food and feed, or cultivation (European Commission 2003a). The method is first verified by the EU-RL GMFF and having completed this step, the EU-RL GMFF organizes an interlaboratory validation to determine if the method satisfies the minimal requirements for reliable GMO testing (CRL-GMFF 2008; European Commission 2003a, 2004c). Therefore, fully validated methods are available for implementation in routine diagnostics laboratories for specific detection and quantification of all GM events approved in the European Union. In contrast, the choice of methods used for screening the presence of GMOs and identification of EU unauthorized GMOs remains the subject of individual research and choice of testing laboratories (Morisset et al. 2009). See the section, Methods.

The implementation of a method for GMO diagnostics is composed of successive stages. The progression from one stage to the next depends on the demonstration of acceptable method performance in the previous stage (Fig. 11):

- Development of the method.
- In-house validation: evaluation of the method performance in a single laboratory, usually the one that has developed the method.
- Prevalidation: the transfer of the method to a limited number of laboratories (usually two to four), for the evaluation of the method performance in other laboratory settings (Codex Committee on Methods of Analysis and Sampling 2010).
- Full validation (also termed interlaboratory method performance study, collaborative study, collaborative trial, or ring trial): a minimum number of laboratories evaluates the method performance according to an internationally accepted protocol (CRL-GMFF 2008). In the following, we use only the term "validation" for this stage of method implementation.
- Verification of the method: before using a new detection method for routine GMO testing, each laboratory has to verify that it can properly use the method for its intended purpose (International Organization for Standardization 2005d).
- Use of the method for routine GMO detection.

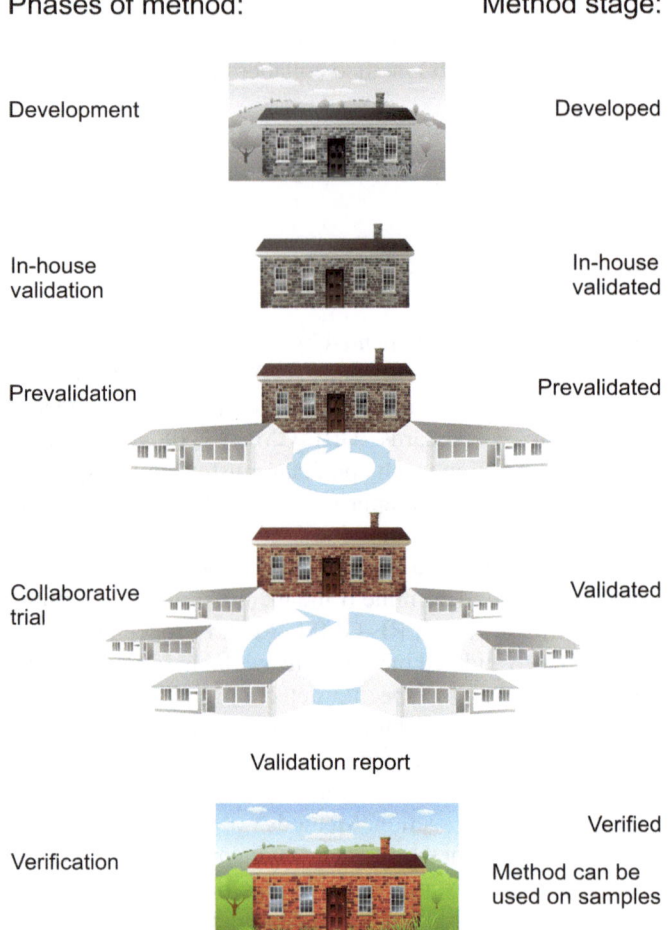

Fig. 11 Implementation of a GMO detection method

The aim of this section is to present a practical approach for the verification of interlaboratory validated methods for the qualitative and quantitative detection of GMOs, including screening, and event-specific methods; the validation of methods not being the main scope of this chapter as well as of this book. Considering that qPCR is the method of choice in the European Union for the identification and quantification of GMOs, we exclusively refer to the system set for verification of this methodology.

The following approach is based on the authors' long experience in GMO diagnostics and is the result of multiple improvements and technical discussions with internal and external experts in the fields of molecular analysis and quality management. This practical approach is exemplified through the verification process of the DP-098140-6 maize event-specific qPCR method.

Method Validation

The goal of the validation is to evaluate the performance characteristics and limitations of an analytical method (Codex Committee on Methods of Analysis and Sampling 2010). During validation, a minimum number of laboratories (12 to 15, usually) participates in an interlaboratory study organized in accordance with internationally accepted requirements, typically the ones described in ISO 5725 (International Organization for Standardization 1994). Further information regarding the requirements and organization of validation of PCR and qPCR methods in the frame of GMO detection can be found in the guidance documents from the EU-RL GMFF (CRL-GMFF 2008; European Commission 2004c), and from the CCMAS (Codex Committee on Methods of Analysis and Sampling 2010), as well as in scientific publications (Taverniers et al. 2004; Žel et al. 2008).

When the outcome of the validation is positive (the performance characteristics satisfy all performance criteria set prior to the validation), the method is considered to be reliable for its purpose. The results of a validation process are usually published in a report or peer-reviewed article and describe which analytes can be determined in what kind of samples in which experimental conditions (Codex Committee on Methods of Analysis and Sampling 2010). Examples of validation reports of the methods proposed for GMO detection can be seen on the EU-RL GMFF website (http://gmo-crl.jrc.ec.europa.eu/). Ideally, the method that was validated according to internationally recognized protocols can be submitted for evaluation to standardization bodies (e.g., the International Organization for Standardization, the European Committee for Standardization) in order to be accepted as a standard method.

Aspects of Modular Approach in Method Verification

Before using the method for routine analysis and even if the detection method is a standard, the laboratory accredited according to the EN ISO/IEC 17025 standard has to provide objective evidence that the performance parameters specified in the validated method have been met (International Organization for Standardization 2005d). Despite the fact that several guidelines on method verification have been published (Thompson et al. 2002; Weitzel et al. 2007) and some publications intend to propose an improved harmonization of this process (Ciabatti et al. 2006; Scholtens et al. 2010; Žel et al. 2008), there is a need for further discussions and room for improvement. Recently a working group of ENGL produced a guidance document on the verification of analytical methods for GMO testing when implementing interlaboratory validated methods (European Network of GMO Laboratories 2011). This ENGL guidance document has been published as a "JRC Scientific and Technical Report from the JRC" (EUR24790 EN), and, as stated earlier, is available on the JRC website at http://jrc.ec.europa.eu/. Although this document is an excellent starting point to get basic information on method verification, it does not cover all aspects of method verification and is planned to be further developed by the working group.

Analytical procedures for detecting GMOs are composed of successive steps including grinding of samples, extraction of DNA, and multiplication of specific targets by PCR. The method verification procedure always entails striking a balance among costs, risks, and technical possibilities (International Organization for Standardization 2005d). In GMO detection, there is a constant need for implementation of methods detecting new GMOs coming on the market and new screening methods, as well as DNA extraction procedures for coping with a wide range of sample types. Therefore, a modular approach of verification, involving a combination of verification of individual analytical steps (termed "modules"), is widely recognized as appropriate. In the area of food and feed testing it is impossible to precisely define sample specificities. Therefore, strict quality controls need to be introduced at both the DNA extraction module and qPCR module levels (Cankar et al. 2006). Introducing such controls ensures the independence of the analytical modules. Such a modular approach leads to reasonable accuracy with substantially reduced costs, as compared with a nonmodular approach, where only the whole procedure (e.g., one extraction together with one qPCR method) should be verified and combinations of individual modules are not possible (Holst-Jensen and Berdal 2004). The general acceptance of modularity was further supported also by the study from Bellochi and coworkers testing the interaction between analytical modules (Bellocchi et al. 2010b).

Strategy for Selecting the Methods to Be Implemented

Currently (September 2011), there are 38 individual event-specific methods fully validated by the EU-RL GMFF and the ENGL, and 38 other methods are in the pipeline (ornamental GM plants were not included; EU-RL GMFF 2011). There are also other published validated methods as recently listed by Kodama and collaborators (Kodama et al. 2010). One should also add the different taxon-specific methods already validated or in the pipeline, several taxon-specific methods existing for the same crop. In addition to event-specific methods, individual laboratories also need to implement screening methods that cover as many GMOs as possible at the same time. The appearance of unauthorized GMOs on the market is also an incentive for rapid implementation of new methods (Žel et al. 2008). Finally, the diversity of reagents (and of course of reference materials) required to perform all these methods adds another level of complexity to the implementation of the methods. Verifying and maintaining all the methods in a laboratory can be a real burden in terms of cost and organization for a laboratory accredited according to the EN ISO/ICE 17025 standard. It is probable that laboratories involved in GMO detection will not implement all these methods and therefore, the implementation of new GMO detection methods relies on the method selection strategy these laboratories follow. It is important to discuss with customers (competent authorities, inspection services, companies) their specific needs to put priorities on which methods to implement and adopt plans accordingly. This strategy can also be influenced by other factors such as the availability of experienced personnel, technical and financial resources (to verify, maintain, and routinely use the methods and associated materials including reference material), the scope of the laboratory (sample types such as seeds, food, feed, environmental samples, etc.), and the availability of reference materials and proficiency tests (Žel et al. 2008).

Verification Procedure

Procedures and responsibilities for development, implementation, and verification of methods should be described in detail in the quality documentation (EA-EUROLAB-EURACHEM Permanent Liaison Group 2001). First, a draft of the internal standard operating procedure (SOP) for the method to be implemented and routinely used should be written. Ideally, the SOP to be used in the laboratory will be exactly the same as the SOP described in the method validation report. However, several factors can lead a laboratory to deviate from the SOP produced during validation.

- For example, five different maize-specific systems targeting two different taxon-specific sequences (the *hmgA*, the alcohol dehydrogenase 1 gene *adh1*) have been validated by the EU-RL GMFF together with ENGL (EU-RL GMFF 2011). To avoid the maintenance of multiple taxon-specific methods and according to the notion of modularity (Holst-Jensen and Berdal 2004), a laboratory can choose to use only one of them for each taxon. As an example, a laboratory that would have historically implemented the *adh1* system from the MON-00603-6 – specific sweet maize event (EU-RL GMFF 2011), could choose to use this system uniquely in parallel with the relevant event-specific method for the quantification of all maize events. Therefore, the laboratory should adapt the SOP of the validated method in order to demonstrate it performs correctly with its preferred taxon-specific system.
- Another factor of adaptation can be the qPCR reagents to be used in the procedure. It may be easier for a laboratory to handle only one type of polymerase and other qPCR reagents for all the detection methods. However, numerous types of qPCR reagents and polymerases have been validated by the EU-RL GMFF and ENGL (EU-RL GMFF 2011). For example, a laboratory may have chosen to use only the universal polymerase and qPCR reagents kit from the SOP of the validated MON-00603-6 -specific method. Therefore, the laboratory should adapt the SOP providing it was demonstrated the assay performs correctly with its preferred polymerase and qPCR reagents during the verification procedure.
- The type of instruments available in the laboratory can be another factor of variation for the SOP to be verified. For example, setting the baseline and threshold as indicated in the validated A5547 specific method (EU-RL GMFF 2011) may not be possible with all qPCR instruments. Therefore, the laboratory should adapt the validated SOP and demonstrate that assay performs correctly with the available qPCR instrument(s).
- Similarly, the reaction volume may need to be modified. So far, most of the qPCR methods validated by the EU-RL GMFF and ENGL need to be performed

in 25-µl reaction volumes. However, some qPCR instruments may not allow these reaction volumes (e.g., 384-well or microfluidics systems as the Fluidigm BioMark instruments). Therefore, the laboratory should adapt the SOP of the validated method and demonstrate it performs correctly in a different reaction volume.

If the laboratory decides to deviate from the SOP of the validated method, this deviation should be well tested during verification of the method. Taking into account all these considerations, a draft version of the internal SOP is written.

Verification Plan

The verification method should follow a precise experimental plan. As an example, after the draft SOP has been written and introduced in the quality system (marked, e.g., as "In Verification" to explicitly specify that this is not yet the working SOP), a verification plan can be written for each method to be implemented, listing the individual parameters to be tested, specifying how these parameters will be experimentally verified, which operators will perform the verification, and any additional details for precise realization of the verification such as the reference material(s) and the equipment to be used. The verification procedure can be organized in such a way that several parameters are tested in the same qPCR run. An example is given in Table 5.

Table 5 Example of verification plan for the qPCR event-specific method for the DP-098140-6 maize line previously validated by EU-RL GMFF. For explanation of individual parameters see text

General information	
Is the method already validated in an interlaboratory study (e.g., by EU-RL GMFF? If so, quote the reference of the validation).	YES. Event-Specific Method for the Quantitation of Maize Line 98140 Using qPCR Protocol; 07.01.2011; http://gmo-crl.jrc.ec.europa.eu/statusofdoss.htm
Reference material ID, GM content? Is reference material available? What kind of reference material is it? (Certified or not? What kind of certificate, characteristics? Is reference material appropriate for validation?).	Certified reference material available at the European Commission – Joint Research Centre – Institute for Reference Materials and Measurements. GM% = 10.00%. The material is heterozygous according to the certification report. Reference material is satisfactory.
Define the qPCR machines to be used and in which tasks they will be used.	Preliminary run and robustness on primers and probes will be done on the Roche light cycler 480 apparatus. Trueness will be done on the ABI 7900HT apparatus. Intermediate precision will be done on the ABI 7900 apparatus and ABI 7900HT.

(continued)

Table 5 (continued)

Preliminary run *(define the concentration(s) of DNA in ng per qPCR reaction)*
100, 10, 2.5, 1.0, 0.5, 0.1, 0.01, and 0.001 ng per qPCR reaction of 10% reference material
DP-098140-6 (IRMM).

Robustness

Different primer concentrations (define the primer concentrations as well as the DNA concentrations in ng).	It is planned to use primers at 500 nM (final concentration). Test how the method works at the following primer concentrations (+/– 20%): 400 nM, 500 nM, 600 nM DNA concentration in the qPCR reaction: 2.5 ng and 0.5 ng.
Different probe concentrations (define the probe concentration as well as the DNA concentrations in ng).	It is planned to use probe at 150 nM (final concentration). Test how the method works at the following probe concentrations (+/– 20%): 160 nM, 200 nM, 240 nM DNA concentration in the qPCR reaction: 2.5 ng and 0.5 ng.

Intermediate precision

The same operator repeats the procedure at different dates (at least two; define operator and the number of repeats).	OPERATOR NAME. Two repeats

Define the concentrations of DNA in ng and copy number used to determine the limit of detection (LOD), limit of quantification (LOQ), the dynamic range, and the threshold Cq value.

From the stock solution of the reference material (ID174), prepare DNA dilutions containing: 100, 10, 2.5, 1.0, 0.5, 0.1, 0.01, and 0.001 ng per qPCR reaction of 10% reference material DP-098140-6 (IRMM). Considering that the material is heterozygous for the transgene in the transformed event (IRMM) and the size of the maize genome,* this corresponds to approximately 1831, 183, 46, 18, 9, 2, 0.2, 0.02 copies of the specific amplicon for DP-098140-6.

Trueness

Internal verification using sample with known content (if no proficiency test available): define the sample.	The ERM reference material ERM- BF427C (2%) could be used. To prepare a material close to 0.1%, this reference material could be spiked to 0.09%.
Participation in proficiency test.	No proficiency test available in the immediate future.

Specificity/selectivity

If the method wasn't already validated in an interlaboratory study, for example, by EU-RL GMFF, define which analyses are needed to be carried out.	Not needed. Method validated in interlaboratory trial.

*In 100 ng of maize DNA there are 18,315 genome copies (2n)

The validation status of a method is an important point to check while preparing the verification plan. If the method to be verified has been validated according to an internationally recognized collaborative study protocol, some performance parameters such as the specificity may not need to be tested during the verification. If specificity was not thoroughly checked in the previous phases of the method implementation, specificity will need to be tested during the verification. This currently applies especially to the so-called screening methods: their specificity may have been properly checked but, as new GM events are regularly introduced in the market, the specificity will need to be verified on these new events.

Another important point while preparing the verification plan is the type of available reference material. In general, a CRM for which sufficient information on its quality and origin are available will be preferred to reference material without a certificate or with a poorly informative certificate. If the certificate is not available, as much information as possible should be gathered: its availability, suitability for qualitative or quantitative analysis, and most important, its level of zygosity. As demonstrated recently, zygosity of a plant reference material can have a strong influence on the final results of quantification especially when using hemizygous material to build a standard curve (Zhang et al. 2008a). See the section, Organization of the Laboratory and Quality Management System.

If more qPCR apparatus is used in the routine analysis, demonstration of the robustness of the method to equipment is also required. It is recommended to assign different qPCR apparatus randomly for the different steps of experimental verification (preliminary run, robustness run, repeatability runs, and trueness run).

Experimental Verification

During the method verification process, all experimental data need to be documented. The selection of parameters to be tested depends on the data obtained from method validation. It is also important to consider the intended application of the method, for example, qualitative or quantitative. As a general rule, methods suitable for quantitative purpose are also suitable for qualitative purpose. Testing conditions during verification (such as reaction volume, PCR machine, etc.) should be the same as when later, the method will be used for the routine analysis.

As described in the above verification plan (Table 5), the experimental verification starts with a preliminary run. If the results from preliminary runs confirm the choice of reference material dilutions indicated in the verification plan, the following experimental steps can be performed in the indicated sequence: "robustness of primers and probe" run, "repeatability" runs, "trueness" run, and "specificity" run (if required). Between each step, the analyst responsible for the verification method should analyze the intermediate results and decide either to proceed to the next step, to repeat the previous experimental step, or to re-evaluate the verification plan and/or the procedure. Performance of each parameter tested during method verification is evaluated against an acceptance criterion. A performance criterion is a specified measure employed in assessing the ability of the assay to perform its intended function (e.g., the minimal sensitivity acceptable to reliably detect a GMO target sequence).

At the end of this section, a practical procedure for verification of a qualitative qPCR method (Table 6) and a quantitative qPCR method (Table 7) is presented. Additionally parameters to be examined during verification of a newly implemented method are summarized (Table 8).

Table 6 Practical procedure for verification of qualitative qPCR method

	DNA concentrations	Number of different DNA concentrations	Number of parallels	Number of repetitions
1. Preliminary test to define appropriate DNA concentrations (and test new chemicals)	In the range of 100 ng – 0 ng	At least 1	2	1
2. Robustness	According to preliminary run	At least 2	3 for every concent-ration	1
3. LOD, intermediate precision, amplification efficiency, and R^2	According to preliminary run	At least 5	3 for every concent-ration	2 – in different days
4. Trueness	Participation in the proficiency test when possible. Otherwise check trueness in an intralaboratory test on a sample with known presence of target analyte.			
5. Specificity/selectivity (if not determined before)	Test a broad range of available target and nontarget material, GMOs, and non-GMOs.			

Table 7 Practical procedure for verification of quantitative qPCR method

	DNA concentrations	Number of different DNA concentrations	Number of parallels	Number of repetitions
1. Preliminary test to define appropriate DNA concentrations (and test new chemicals)	In the range of 100 ng – 0 ng	At least 5	2	1
2. Robustness	According to preliminary run: One clearly above expected LOQ; one between expected LOD–LOQ	At least 2	5 for every concentr-ation	1
3. LOD, LOQ, intermediate precision, amplification efficiency, and R^2	According to preliminary run	At least 7 (at least 2 should be around LOD and 2 around LOQ)	5 for every concentr-ation	2 – in differ-ent days
4. Trueness	An intralaboratory test on an independent sample with known% of target analyte and/or participation in a proficiency test.			
5. Specificity/selectivity (if not determined before)	Test a broad range of available target and nontarget material, GMOs and non-GMOs.			

Table 8 Parameters to be examined during verification of a newly implemented method

Parameters	Implementation of method with changed key parameters*	Implementing a method validated in interlaboratory trial**
All methods		
Robustness	YES	YES
Intermediate precision, amplification efficiency, and R^2	YES	YES
LOD	YES	YES
Trueness	YES	YES
Specificity/selectivity	YES	NO
Quantitative methods only		
LOQ	YES	YES
Dynamic range	YES	YES

*If laboratory decides to deviate from the SOP of a method validated in interlaboratory trial
**For example, by the EU-RL GMFF and the ENGL

Preliminary Run

Definition: This run is purely informative as it is performed in order to predetermine the DNA dilutions that contain target copies close to the limit of detection (LOD). It also usually helps define dilutions in the quantitative range of the method to be used in the next verification experimental steps.

Procedure: At least five different concentrations of the reference material in the range 25 ng/μl–0 ng/μl are tested in two parallels. These dilutions can be the same planned for robustness and intermediate precision runs. If possible, two of these dilutions should contain a low target copy number, close to the expected LOD.

Acceptance criterion: None.

Actions: Calculate appropriate range of sample dilutions for next verification steps.

Robustness

Definition: Measure of a method capacity to remain unaffected by small, but deliberate deviations from the experimental conditions described in the procedure (CRL-GMFF 2008).

Remark: If the method procedure for verification is exactly the same as the validated method (including the equipment, reference material, and reagents), the robustness does not need to be verified. If the procedure deviates from the validated method (reaction volume, cycling parameters, reagent nature, and/or concentration, etc.) robustness shall be verified.

Procedure: Robustness is tested by varying, for example, the concentrations of primers and the probe. A 20% variation of the primers and probe concentrations of the validated method is tested as this can be pipetting error that can realistically be expected. Robustness is determined on the reference material at two DNA concentrations.

Quantitative purpose: One DNA concentration is chosen near the expected LOD (according to the preliminary run), and one within the expected linear range of amplification (according to the preliminary run). For each primer and probe concentration at each DNA concentration, the measurement is done in five replicates.

Qualitative purpose: One DNA concentration is chosen near the expected LOD (according to the preliminary run), and one at higher target concentration, around 50 copies or above (according to the preliminary run). For each primer and probe concentration at each DNA concentration, the measurement is done in three replicates.

Acceptance criterion:

Quantitative purpose: The method is robust if despite varied parameters the determined analyte copy number does not change more than 30%. (See Real-Time PCR for calculation of analyte copy numbers.)

Qualitative purpose: The method is robust if despite varied parameters, presence of the target is still detected.

Intermediate Precision

The precision of an analytical procedure expresses the closeness of agreement between a series of measurements obtained from multiple replicates of the same homogeneous sample under the prescribed conditions. Intermediate precision may be measured by altering the following variables: different days, different operators, and different equipment (Weitzel et al. 2007). Quantitative measures of precision depend critically on the stipulated conditions. Repeatability and reproducibility are particular sets of extreme conditions (EURACHEM 1998). Repeatability conditions are conditions where test results are obtained with the same method, on identical test items, in the same laboratory, by the same operator, using the same equipment within short intervals of time (CRL-GMFF 2008). If precision is determined under repeatability conditions, precision data can be expressed as relative repeatability standard deviation (RSDr). We use several types of machines so we do not strictly assess repeatability of the method. Therefore, in our case, RSD is the relative standard deviation of test results obtained under intermediate precision conditions.

Procedure: Intermediate precision is calculated from replicate qPCR runs and is carried out on at least two different days by the same operator or, if possible, by an additional operator.

Quantitative purpose: At least seven different DNA concentrations of the reference material are analyzed in the range of 100–0 ng/µl determined according to the preliminary run. At least two concentrations should be close to the expected LOD, and at least two concentrations should be close to the limit of quantification (LOQ; according to the preliminary run). The procedure is carried out in five parallels for each DNA concentration.

Qualitative purpose: At least five different DNA concentrations of the reference material are analyzed in the range of 100–0 ng/µl determined according to the preliminary run. At least two concentrations should be close to the expected LOD (according to the preliminary run). The procedure is carried out in three parallels for each DNA concentration.

Remark: If after several method verifications it can be statistically demonstrated that the variability between operators during intermediate precision tests is negligible, the intermediate precision tests can be performed by only one operator on two different days. Such well-documented observations can contribute to quicker and less costly verification schemes (Žel et al. 2008).

Acceptance criterion:

Quantitative purpose: The method is acceptable if the RSD of the determined analyte copy number is ≤25% over the whole dynamic range of the method (CRL-GMFF 2008).

Qualitative purpose: The method is acceptable if the results in terms of positive/negative signals are comparable between the runs. Additionally the amplification efficiency must be within the limits described below (see Amplification Efficiency).

Limit of Detection (LOD)

Definition: The lowest amount or concentration of analyte in a sample that can be detected reliably but not necessarily quantified (International Organization for Standardization 2006). Experimentally, methods should detect the presence of the analyte at least 95% of the times at the LOD, ensuring ≤5% false negative results. In our experience it is most appropriate to focus on the absolute LOD, expressed in target copy number as the relative LOD of the method (smallest detectable %GMO – ratio transgene/endogene that can be reliably detected) also relies on the absolute LOD of the same method (smallest number of transgene copies that can be reliably detected). If the method is proven to reliably detect low target content in a high DNA background, the limiting factor for sensitivity of the method is only the absolute target copy number.

Procedure: The assessment of the LOD is done on the data generated during the intermediate precision runs. To calculate the absolute LOD of a method with 95% confidence, it is theoretically necessary to analyze 60 qPCR replicates for each tested concentration. As this is not very practical, our laboratory uses the term false negative rate on a lower number of replicates, such as ten replicates. The definition of false negative rate is the probability that a known positive test sample is classified as negative by the method. An increase in the false negative rate is observed when the amount of analyte approaches the LOD of the method. The false negative rate has to be below 5% (i.e., all ten qPCR replicates have to be positive; European Network of GMO Laboratories 2011). The LOD of the method is therefore the number of target copies corresponding to the last dilution for which all test parallels (in our example ten qPCR replicates) give positive results.

Acceptance criterion: In our laboratory we have set the LOD acceptance criteria at 25 copies of analyte DNA. Assuming a Poisson distribution of target copies, the probability distribution suggests that a single target copy in a qPCR reaction should give approximately 30% negatives. Therefore, at least one replicate has to be negative for a DNA level estimated to be at one or less target copy (and therefore below LOD; Bustin et al. 2009). If this last condition is not fulfilled the DNA concentration must be re-estimated.

Threshold Cq Value

Definition: The last Cq value considered as a positive result when using the verified method. It is usually observed that there is a large uncertainty when measuring the concentration of DNA to be used in qPCR reactions. Also, the intensity threshold set when analyzing qPCR results can vary considerably between two qPCR runs, two operators, and even two thermocyclers. Therefore, the Cq value corresponding to the LOD when verifying the method cannot be used as a threshold to decide whether the target is present in the sample.

Procedure: Cq values obtained during intermediate precision runs of the method verification are carefully checked. In the area close to LOD, where some wells start giving "undetermined" results, the range of the highest Cq values observed (not considering outliers) is used. This Cq value is then rounded to the upper half Cq value and another 0.5 Cq value is added to take into account the difference in threshold chosen between runs.

In the case of the screening methods, the threshold Cq value should be checked on every GMO event available in the laboratory at the time of the method verification. For this, five parallel reactions of five dilutions around the LOD should be performed for each tested GMO event. Once the screening method is verified, reference material for new GMO events may become available in the laboratory and should be checked for the threshold Cq value with the screening method. If this threshold Cq value is higher than the previous one, this value should be updated accordingly.

Acceptance criterion: None.

Limit of Quantification (LOQ)

Definition: The lowest amount or concentration of analyte in a sample that can be quantitatively determined with an acceptable level of precision and accuracy (International Organization for Standardization 2006). In our experience it is most appropriate to focus on the absolute LOQ, expressed in target copy number as the relative LOQ of the method (smallest ratio transgene/endogene that can be reliably quantified) also relies on the absolute LOQ of the method (smallest number of transgene copies that can be reliably quantified). If the method is proven to quantify reliably low target content in a high DNA background, the limiting factor for quantification of the method is only the absolute target copy number.

Remark: LOQ does not need to be verified if the method is used for qualitative purposes.

Procedure: The assessment of the LOQ is done on the data generated during the intermediate precision runs. The LOQ of the method is the lowest number of target copies for which the RSD of the determined analyte copy number is ≤25%.

Acceptance criterion: In our laboratory we have set the LOQ acceptance criterion at 50 copies of analyte DNA.

Dynamic Range

Definition: The range of concentrations over which the method performs in a linear manner with an acceptable level of trueness and precision (CRL-GMFF 2008).

Procedure: The assessment of the dynamic range is done on the data generated during the intermediate precision runs. Within the dynamic range the RSD for the analyte copy number between the parallels must not exceed 25%. In addition the efficiency of amplification must be inside the limits described below (see Amplification Efficiency).

Acceptance criterion: The method's dynamic range must extend from the LOQ to 50 ng of DNA per reaction. In the case where the highest DNA concentration prepared from the reference material is below 50 ng per reaction, the target DNA can be spiked with nontarget DNA to reach this concentration.

R^2 Coefficient

Definition: The coefficient of determination calculated as the square of the correlation coefficient (between the measured Cq value and the logarithm of the concentration) of a standard curve obtained by linear regression analysis (European Network of GMO Laboratories 2011).

Procedure: The assessment of the R^2 coefficient is done on the data generated during the intermediate precision runs.

Acceptance criterion: The average value of R^2 must be ≥ 0.98 through the dynamic range of amplification (CRL-GMFF 2008).

Amplification Efficiency

Definition: The rate of PCR amplification that leads to a theoretical slope of -3.32 corresponds to

$$\text{Efficiency} = (10^{(-1/slope)} - 1) \times 100\% \text{ (CRL} - \text{GMFF 2008)}$$

Procedure: The efficiency assessment is done on the data generated during the intermediate precision runs.

Acceptance criterion:

Quantitative purpose: The average value of the slope of the standard curve must be in the range of $(-3.1 \geq \text{slope} \geq -3.6)$ along the dynamic range of amplification (CRL-GMFF 2008). This corresponds to an amplification efficiency of 90% to 110%.

Qualitative purpose: The average value of the slope of the standard curve must be in the range of $(-2.9 \geq \text{slope} \geq -4.1)$ along the dynamic range of amplification. This criterion is less stringent than for quantitative purpose because it should not affect the final result in qualitative terms (presence/absence of target).

Trueness

Definition: The closeness of agreement between the average values obtained from a large series of test results and an accepted reference value. By definition, it applies only to quantitative methods. The measure of trueness is usually expressed in terms of bias (CRL-GMFF 2008).

Procedure: Trueness is tested on independent reference material (different from the one used in the other steps of verification) with known content of the target DNA. If the independent sample with known presence of target analyte is not available, spiking of the reference material can be performed to generate an artificial reference material. Trueness of the method can be additionally confirmed by later participation of the laboratory in proficiency testing programs where a Z-score can provide an indication of the trueness of the method. Participation in proficiency tests can be also an alternative to this step of method verification when the CRMs for estimating the trueness are not available. The trueness is measured on the CRM, on at least two different concentrations. One of these concentrations should be close to the legal LOQ (e.g., 0.1%), and one at an upper concentration (e.g., 5%). Alternatively, a reference sample can be made from a higher percentage CRM by spiking it with background DNA.

Acceptance criterion: The trueness shall be within $\pm 25\%$ of the accepted reference value or a Z-score within the range of 2 and -2 should be obtained.

Specificity and Selectivity

Definitions: Selectivity of a method refers to the extent to which it can determine particular analytes under given conditions in mixtures or samples, simple or complex, without interference from other components (Vessman et al. 2001). In contrast with selectivity, specificity is the property of the method to respond exclusively to the characteristic or the analyte of interest (CRL-GMFF 2008).

Remark: Specificity/selectivity should be part of method development and validation; this parameter may not need to be repeated when verifying the method.

Procedure: The specificity/selectivity should be checked on reference material for every GM and non-GM line available to the laboratory, including the ones for which the targeted DNA sequence is present.

For all methods: The presence or absence of the analyte has to be properly determined.

For screening methods: Once the method has been verified, its specificity should be tested on reference material for all new GM lines made available in the laboratory. It is important to be aware that CRMs are certified for the presence of a given event and not for the absence of other events.

For taxon-specific methods: Once the method has been verified, its specificity should be tested on reference material for all new taxons made available in the laboratory.

Acceptance criterion: Only the target of interest is detected with the method.

For event-specific methods: The method must be specific (positive result) to the tested event and must not give positive results for any other DNA sequence, be it found in GMO or not.

For taxon-specific methods: The method must detect the taxon-specific gene for a given plant taxon, and must not give positive results for any other taxon or DNA sequences. If the method is for quantitative purposes it must be verified that the constant number of copies of the taxon-specific gene is present in the different varieties. If possible it should be present in a single copy.

For screening methods: The method must detect the specific genetic element in all test samples harboring the target, and it must not give positive results for any other DNA sequences.

Data Administration

Basic data about the results of the method verification, the report produced during the verification, and the conclusions about the suitability of the method must be written in a document linked to the verification plan. All details regarding the verification procedure and its results must be included in a report of the implemented method.

The conclusion about method applicability and any deviation from the accepted procedure and verification plan have to be documented in this report.

Once the method has been verified, a final version of the SOP should be written. This document will describe the method that has been verified, its intended use, and its experimental procedure. Only the methods that have fulfilled the verification acceptance criteria can be used for further analyses on samples and if the laboratory is accredited they can be added to the list of accredited methods.

Homogenization and DNA Extraction

Introduction

Sample homogenization and extraction of DNA are extremely important steps in the detection of GMOs (Fig. 12) as significant analytical errors can be introduced at this stage. Homogenization is required for two reasons: to achieve

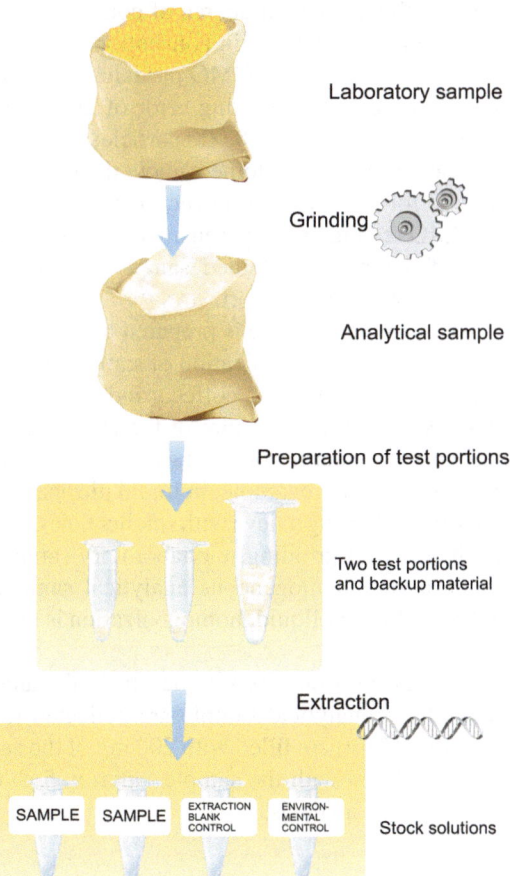

Fig. 12 Work flow of the sample showing the process from homogenization of the laboratory sample, preparation of test portions, extraction of DNA, and finally acquisition of stock solutions of DNA for further analysis

sufficient efficiency of DNA extraction and, above all, to ensure homogeneity and equal representation of GMO-derived particles in the subsamples (International Organization for Standardization 2005c). The extraction method must provide DNA suitable for subsequent analysis, namely a sufficient amount of DNA that is of appropriate quality and structural integrity (Cankar et al. 2006). Highly fragmented DNA and coextracted impurities of a DNA preparation may hinder the correct detection and quantification of GMO in the sample (CRL-GMFF 2008) and that is why extraction methods should also be evaluated prior to implementation in routine laboratory testing (European Network of GMO Laboratories 2011).

Homogenization

The laboratory sample is the portion of material to be used in the laboratory for the analyses. The vast majority of samples must be homogenized prior to qPCR analysis. The laboratory sample should be of a size that ensures the quantification of GMO with a statistical degree of confidence of 95%. Given the threshold value of 1% of GM material within conventional material (in the case of expected inhomogeneous distribution of GMO particles in the investigated material) and taking into account an overall sampling error of 20%, laboratory samples for GMO analysis should contain at least 10,000 particles (Hubner et al. 2001). The laboratory should take care to inform customers about the proper size of laboratory samples they shall provide for analysis. Moreover, customers should be aware that the methods used for GMO testing are very sensitive, which could lead to false positive results in the case of incorrectly handled samples, for example, possible cross-contamination during storage or transport.

An analytical sample is prepared by grinding the laboratory sample if necessary (Fig. 12). The homogenization of samples may be achieved with mills, homogenizers, immersion blenders, coffee grinders, or a suitable equivalent device, depending on the size and the structure of the laboratory sample. If the homogenization with the chosen grinder is not satisfactory, it is necessary to further homogenize the laboratory sample, for example, in liquid nitrogen with mortar and pestle, rehomogenizing in the mill using a sieve with smaller holes, or changing the type of homogenization machine. After grinding, the laboratory sample is once more thoroughly mixed to obtain a very homogeneous analytical sample. In the case where the laboratory sample is flour or liquid, homogenization is not needed but mixing or shaking is still necessary.

Two test portions are sampled from the analytical sample (Fig. 12). Small portions of the analytical sample are collected randomly and weighed; for example, two tubes (2 ml) are filled with 200 mg of the sample, each. An additional tube (e.g., 15 ml) is filled with the sample (approx. 5 g) to be stored as back-up for possible repeats of the analysis.

Risk of Cross-Contamination

In all analytical steps it is necessary to avoid any cross-contamination. Homogenization is the step with the highest contamination risk. During grinding, a fine dust often appears that could contaminate subsequent laboratory samples. As described in the section Organization of the Laboratory and Quality Management System, grinding is kept well separated from all other analytical steps. During homogenization, other precautionary actions should be taken: There is always only one laboratory sample in the procedure. After the homogenization of each sample, laboratory coats and gloves are replaced. Tools and equipment that are used in the homogenization and weighing are washed with tap water after each sample, dried, then wiped with a solution for the removal of DNA and wiped again with 70% ethanol or distilled water. Tubes with test portions are cleaned on the outer surface before they are transferred to the following analytical step.

DNA Extraction

As has already been demonstrated, extraction techniques and sample properties have important impacts on DNA quantification (Cankar et al. 2006). The aim of an extraction method is to obtain an appropriate amount of DNA that is also of sufficient integrity as well as free of coextracted substances that can influence amplification (e.g., phenols, lipids, polysaccharides, high ion concentrations). Extraction of DNA from grains, oilseeds, fruits, and other parts of plants usually gives DNA of appropriate quality, whereas extraction from highly processed food or feed is more challenging. Several methods are available for DNA extraction suitable for different types of samples. Most commonly used are commercially available kits and detergent-based methods (Demeke and Jenkins 2009). Generally when kits are used, there is little possibility to change individual components of the extraction procedure, especially as most manufacturers do not provide sufficient information on the constituents of kit components. In our laboratory, we have implemented five different extraction methods: four kits and one detergent-based method during the 10-year period of routine GMO testing. However, lately we have mostly been using the commercial NucleoSpin food kit (Macherey–Nagel) for almost all samples. We use the cetyltrimethylammonium bromide (CTAB) method (International Organization for Standardization 2005c) for some samples or as an alternative method in the case of unsuccessful extraction and some other protocols where there are special sample types.

List of Tested Samples

Throughout our long experience of routine testing, we have extracted DNA from many different samples. The information on efficiency of extraction methods for certain types of sample is recorded in a list of tested samples. We also include information on any modifications of the standard extraction protocol. This is very useful and helps the operator to choose the most suitable extraction method for the sample in the analysis. If extraction is to be done on a new type of sample, we usually extract DNA with the method that shows the best performance on the majority of samples. If this method fails, another method is used that has been successful in extraction of DNA on similar samples (e.g., high-fat, polysaccharides, proteins, highly processed) based on data from the list of tested samples.

DNA Extraction from Samples

DNA is extracted from each sample in two parallels (Fig. 12). When a commercial kit is used, the protocol for DNA extraction is generally set following the user's manual. Some samples, however, have special characteristics; therefore adaptation of the basic procedure is necessary (e.g., more lysis buffer is used for powdered hygroscopic samples, lower elution volume is used for samples with low content of DNA). The final step in DNA extraction is elution from the binding surface. To obtain better yields, two sequential elution steps are always performed in our laboratory.

Based on the laboratory experience with DNA extractions from particular sample types (DNA quantity obtained), recommendations can be set for the most appropriate elution volumes and subsequent dilutions of stock solutions that are most proper for further qPCR analysis. In our case, if DNA is extracted from maize grain, the elution volume is 100 µl in each elution step and this is the stock solution. In DNA extraction from soya bean grain, the same elution volume 2×100 µl is immediately diluted 1.5× with water. In both cases, DNA concentration in the stock solution (Fig. 12) is approximately 25 ng/µl, which is optimal for qPCR performance. When DNA is extracted from a new unknown type of sample, a lower elution volume is used (2×50 µl) and DNA is quantified prior to qPCR analysis.

Quality Control

Regardless of the method used, quality controls are always included in the extraction procedure. In EN ISO/IEC 24276 it is specified which controls are mandatory and which are recommended. In our laboratory for each extraction series, we perform extraction blank control and environment control (International Organization for Standardization 2006). The extraction blank control is the negative control of the extraction process. Water is used instead of a sample and all DNA extraction steps are performed as if it were a normal sample.

The extraction blank control is used to demonstrate the absence of sample contamination during extraction and should always be the last sample in the extraction series. If more than ten extractions are done in parallel, an extraction blank control is included after each series of ten extractions.

With environment controls, possible presence of GMO dust in the lab environment is checked. A tube with a volume of water equal to the elution volume of samples is left opened during the DNA extraction procedure. Environment control and extraction blank control are tested with qPCR in parallel with the samples using appropriate taxon-specific methods (see Real-Time PCR).

When a new set of reagents for DNA extraction is used, they are tested using positive extraction control. A sample from which high-quality DNA can be extracted should be used as the positive extraction control. The extraction procedure should be performed as for any other sample.

DNA Quantification

Estimation of DNA concentration is another important step prior to qPCR analysis. In the first years of routine analysis, when we were setting up GMO detection systems, all extracted DNA was quantified using the Picogreen fluorescent dye. When this approach was compared to UV spectrometry and agarose gel electrophoresis (Bellocchi et al. 2010a), it was shown that each method has advantages and limitations similar to the later study by Demeke and Jenkins (2009 ; e.g., fluorimetry is rapid and relatively inexpensive, but tends to underestimate the DNA concentration).

In recent years, we have not been quantifying DNA prior to qPCR. The quantity of DNA is assessed during qPCR analysis, inasmuch as we can estimate the quantity of DNA from Cq values obtained by using taxon-specific methods on samples. Only in special cases, when it is not possible to detect targets (neither screening elements nor taxon-specific sequences), DNA concentration and quality are checked by agarose gel electrophoresis. As described above, we keep records on DNA extraction performance with known samples and on appropriate stock solutions to be prepared.

Real-Time PCR

Introduction

Real-time PCR (qPCR) is by far the most widely used technique for detection, identification, and quantification of GMOs. In spite of the wide acceptance of this approach in GMO testing laboratories, there are some differences in understanding and implementing the individual steps of GMO testing. Detailed differences in the method application, reference materials chosen, and differences in data analysis, interpretation, and reporting can lead to major differences in final test results. In addition, new equipment and chemistries for qPCR are constantly evolving towards improvement in analysis and cost-efficiency. Although there is legislation in place and standards are setting general requirements as well as detailed procedures, each

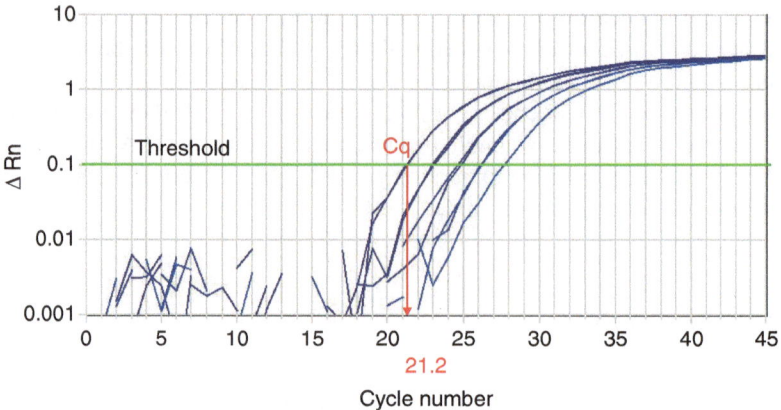

Fig. 13 Graphical representation of qPCR data. Δ*Rn*: fluorescence of the reporter dye divided by the fluorescence of a passive reference dye minus the baseline; Threshold: fluorescence intensity determined as the lowest limit of fluorescence detection. Cq: the intersection between an amplification curve and a threshold, a relative measure of the target concentration in the reaction

laboratory implements the requirements according to its capacities. In this session we show the workflow of routine analysis for quantification of GMO in the laboratory. Besides quantification of GMO content, screening for GMOs and identification of specific lines with event-specific methods are also described (Fig. 13).

Based on several years of experience with DNA extraction and qPCR technology as well as validation and verification of methods, we have organized our testing with the major goal of producing reliable results. Some further adjustments enabled faster and more cost-efficient analysis. In the Homogenization and DNA Extraction section, we already described why we do not quantify the DNA sample before qPCR analysis. However, it is not possible to take a shortcut in the quality control of testing. Every step of the analysis must be controlled and recorded (see the section, Organization of the Laboratory and Quality Management System).

Reporting of results should be harmonized to allow worldwide common understanding of analysis results. Through the activities of the EU-RL GMFF and the ENGL, harmonization of GMO analysis has been achieved in the European Union. However, further work and discussions are still needed to achieve global harmonization on GMO analysis. Results can be expressed as the mass fraction, as the transgene DNA copy number in relation to target taxon-specific sequence copy number calculated in terms of haploid genomes, or as the number of GM seeds ratio. Regardless of the way the results are expressed, the test report must provide information about the uncertainties and limitations associated with the test results (Holst-Jensen 2009). At the end of this chapter, we give an example of a test report. The intention is not to present this as a prescribed document, but to show the key information that should be given to the customer.

qPCR Equipment

Several apparatus are available for qPCR analysis with different properties (e.g., 96-well or 384-well format, excitation with argon-ion laser, LEDs, or xenon lamps, detection with spectrograph and CCD camera, photomultiplier tube, or photodiodes). In routine detection of GMOs, Applied Biosystems machines together with their hydrolysis chemistry procedures are still the most widely used. However, with new developments in the area of qPCR, many different machines are already available on the market and at the moment there are limited data available regarding the comparison of their performance. A small collaborative trial with eight different qPCR machines (Applied Biosystems 7000, 7700, and 7900, Cepheid Smartcycler II, BioRad Chromo4, Corbett Research Rotorgene 3000, and Roche Light Cycler) was organized within the Co-Extra project in the European Union's Sixth Framework Programme (http://www.coextra.eu/). Primers, probes, DNA samples, and CRMs were prepared and distributed as a kit in order to minimize variation due to factors other than equipment. The protocol for the TaqMan® quantification of GM event MON-00810-6 allowed for reagents, volumes, and other conditions to be identical among different machines. The hypothesis of the study was that a large component of the variation of GMO content estimates is due to the type of hardware used. But the results showed that for quantification of GM event MON-00810-6 the effect of the machine used was not significant, if proper conditions were set (e.g., consistent calibrants). Moreover, there was no obvious systematic error or bias associated with any individual machine (Allnutt et al. 2010).

In our laboratory, two different types of machines are used for GMO quantification (Applied Biosystems 7900 and Roche Light Cycler 480). With both types of equipment, we only use 384-well plates. Roche Light Cycler 480 was introduced after the Applied Biosystems 7900 and the transfer to this new machine was carefully checked due to the difference in fluorescence detection and in the analysis of raw data. There were already quantifications of several GMOs verified on ABI 7900. As mentioned in the verification section, the introduction of a new machine requires a verification of method performance. After the introduction of the Roche Light Cycler 480 in our laboratory, we ran one of the international validation studies in parallel on both machines. Comparison of both types of qPCR equipment has shown that the Roche Light Cycler 480 is appropriate for use in routine analysis in our laboratory. Because the decision was only based on quantification of one GMO, further experiments were performed for the transferability of the methods between machines. Intermediate precision, trueness, and robustness of several specific events (MON-00021-9, MON-04032-6, MON88017-3, and MON-89788-1) and taxon-specific sequences (*adh* and *hmgA*) were tested and showed that the Roche Light Cycler 480 machine provides results of comparable quality with the ABI 7900 apparatus, and is thus appropriate for use in routine analysis.

Chemistries

All qPCR systems rely upon the detection and quantification of a fluorescent reporter, the signal of which increases in direct proportion to the amount of PCR product in a reaction. Various chemistries for the generation of fluorescent signal have been developed: intercalating or sequence-unspecific DNA labeling dyes (SYBR®Green), primer-based technologies (AmpliFluor, Plexor, Lux primers), and technologies involving double-labeled probes detecting hybridization of the probe to the target (molecular beacon) and hydrolysis of the probe (TaqMan, CPT, LNA, and MGB) (Buh Gasparic et al. 2010; Weighardt 2006). Although TaqMan and SYBR Green are the most commonly used in routine laboratories, other chemistries are rapidly developing and their performance in GMO detection was recently assessed. Comparative evaluation of performance characteristics such as specificity, sensitivity, intermediate precision, robustness, and dynamic range for nine different qPCR chemistries showed that there is no significant difference, but some may need more optimization prior to use. In addition, there are also differences in practicability and cost effectiveness that may influence the decision on chemistry (Buh Gasparic et al. 2010).

In our laboratory, TaqMan chemistry is mainly used, which is mostly due to the fact that we are implementing methods previously validated by EU-RL GMFF. In spite of evidence that less expensive chemistries are also suitable for GMO detection, transfer to new chemistry would represent enormous labor and costs due to verification that would be necessary for our accredited methods.

qPCR Setup, Analysis, and Interpretation

In order to ensure accurate and reliable results of qPCR methods, a laboratory should follow standards and guidelines mentioned in previous sections (International Organization for Standardization 2005a, b, d, 2006), where general requirements, principles, minimum requirements, and performance criteria are laid down as well as specific qPCR procedure steps. For specific procedures, laboratories can also find information in different GMO detection manuals (European Union Reference Laboratory for GM Food and Feed and European Network of GMO Laboratories 2010; Querci et al. 2004), on Internet pages (e.g., http://gmo-crl.jrc.ec.europa.eu/ or http://gmofmdb1.jrc.it/fmi/iwp/cgi?-db=GMOmethods&-loadframes), or in the literature (European Union Reference Laboratory for GM Food and Feed and European Network of GMO Laboratories 2010; Grohmann et al. 2011).

The following is an example of our approach for the setup of routine sample analysis using qPCR. We do not go into details of procedures, but rather focus on quality controls, data analysis, and interpretation of results. As an example, we follow the procedure for testing maize flour using Applied Biosystems' machine for qPCR analysis and the Excel electronic spreadsheet program for calculations. Our testing starts with the DNA that was extracted in two parallels from two independent test portions (see the section, Homogenization and DNA Extraction).

Qualitative Analysis: Screening

In the first step of routine sample testing, screening analysis is performed in 10-µl reactions. In DNA extracted from maize flour, the presence of two screening elements is checked, namely P-35S and T-NOS. Additionally, the presence of taxon-specific sequence is analyzed mostly for quality control. Different assays are available for the maize-specific sequence. In our case *hmgA* is used. Recently we implemented the pentaplex method, developed by the Bavarian Health and Food Safety Authority (LGL) in Germany (manuscript in preparation for publication) and we expect that in the near future the use of the GMOseek algorithm will ease the choice of screening methods (see the section, New Challenges). Anyway, here we show the example of analyses on two screening elements P-35S and T-NOS, because this is the most common approach in the majority of GMO testing laboratories.

Sample Analysis

As the DNA is extracted from the sample in two parallels and each parallel is tested in duplicate, there are four qPCR reactions for each sample and for each target (both screening elements and taxon-specific sequence). To consider the testing result as positive, more than half (three out of four in this case) of the reactions should be positive. However, negative results are not only those giving signals below the threshold, but also Cq values that are above the predetermined threshold Cq value. As explained in the Verification of Methods section, threshold Cq value for each analyte is determined during verification of the method indicating the Cq value that is still accepted as a positive signal as it should represent amplification of one copy of the target DNA in the reaction. If there is one negative and one positive result within each parallel of the DNA extraction, the result is presented as negative. A testing result is ambiguous when the result of both replicates in one parallel is positive and both replicates in the second parallel result negatively. In that case, it is necessary to repeat the qPCR analysis with an increased quantity of DNA, if possible. If the result is still ambiguous, the whole analysis including the DNA extraction should be repeated. If the result is the same again (one parallel positive, one negative), we consider the analysis result as negative. The same approach is used in all following testing steps.

Quality Controls

Sample testing includes quality controls for DNA extraction (described in the section, Homogenization and DNA Extraction) and for qPCR reactions.

DNA Extraction Controls

Two quality controls that were prepared during extraction of DNA, namely environment control and extraction blank control, are tested only for the taxon-specific

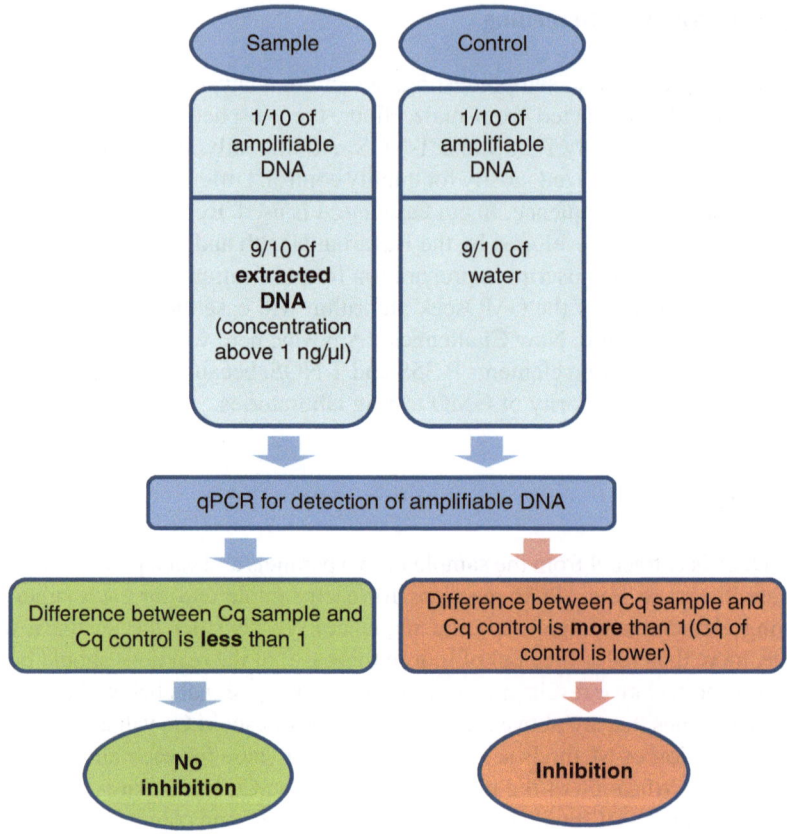

Fig. 14 Spiking can be used to check for the presence of inhibitors (see text for more details)

sequence, in our case example this is the *hmgA* target. If the *hmgA* target is detected in the environment control, contamination of the environment (workplace) during DNA extraction should be considered, whereas in the case of detection of the *hmgA* target in the extraction blank control, contamination of reagents or samples during DNA extraction should be considered. In both cases, results of sample analysis are not reliable: the analysis is not valid and should be repeated, including the extraction of DNA.

In the screening analysis, the taxon-specific sequence is also used as the control of successful extraction. If neither the transgene (P-35S or T-NOS), nor the taxon-specific sequence (*hmgA*) could be detected in the sample, the concentration of extracted DNA should be checked as described in the Homogenization and DNA Extraction section. If concentration of DNA is below 1 ng/μl, it is too low for qPCR analysis. DNA extraction should be repeated with another method, unless this type of sample was already tested and no other suitable extraction method is available. If the concentration is higher than 1 ng/μl, the extracted DNA should be checked for the presence of inhibitors with the DNA spiking approach (Fig. 14). Analysis is repeated on the sample DNA mixed

Table 9 Quality controls of DNA extraction and qPCR reactions. NTC: no template control

Sample name	Cq	Conclusion
hmgA		
NTC	Undetermined	OK
NTC	Undetermined	OK
Positive DNA target control	24.65	OK
Positive DNA target control	24.65	OK
Extraction blank control	Undetermined	OK
Environment control	Undetermined	OK
P-35S		
NTC	Undetermined	OK
NTC	Undetermined	OK
Positive DNA target control	28.78	OK
Positive DNA target control	28.92	OK
T-NOS		
NTC	Undetermined	OK
NTC	Undetermined	OK
Positive DNA target control	25.89	OK
Positive DNA target control	25.76	OK

with 1/10 volume of DNA that was previously successfully amplified without any indication of inhibitors (amplifiable DNA). The results are compared with the control composed of 1/10 volume of amplifiable DNA and 9/10 volume of water. If Cq values of the sample and the ones of the control do not differ by more than one Cq, it means there is no inhibitor in the extracted DNA. In this case, the sample analysis result is negative. A larger difference in Cq values indicates the presence of inhibitors in extracted DNA. DNA extraction must be repeated by the same method (if there is a suspicion of a mistake in the performance of the procedure), by another method (which can lead to the DNA without qPCR inhibitors), or analysis can be finished (if a similar type of sample was already tested and it was not possible to extract DNA by different methods). See also the section, Homogenization and DNA Extraction.

Real-Time PCR Controls

For the control of the qPCR, an amplification reagent control, no template control (NTC), and a positive DNA target control are checked (Table 9). NTC contains all the reagents used for qPCR, however, sterile deionized water is finally added into reaction instead of the test DNA. NTC should be tested for the same analytes as the sample in duplicate, one at the start of pipetting and one at the end for each analyte.

Table 10 For control of inhibition one extraction parallel is analyzed in stock solution and tenfold dilution. ΔCq or slope is calculated. 1×: stock solution, 10×: 10× dilution

hmgA

Sample name	Cq	Relative conc.	Log$_{10}$ Rel. Conc.	Slope	Conclusion
Sample1- parallel1 1×	21.07	10	1	−4.4916	*hmgA* confirmed
Sample1- parallel1 1×	21.34	10	1		weak inhibition
Sample1- parallel1 10×	25.74	1	0		
Sample1- parallel1 10×	25.64	1	0		
Sample1- parallel2 1×	21.65				
Sample1- parallel2 1×	21.46				

NTC's analysis for all amplicons should not result in a positive qPCR signal; otherwise contamination during preparation of qPCR should be considered. A reference material containing the target sequence(s) is used as a positive DNA target control. Moreover, this is the material with the estimated copy number of target sequences, inasmuch as these data are also used for calculation of practical LOD/LOQ, as explained later. The positive DNA target control should be tested in duplicate and for the same analytes as the sample. The presence of all targeted sequences must be confirmed and must be in a predicted range (as set from our experience) otherwise the analysis is not valid. If targeted sequences are not detected, the analysis is also not valid. In both cases, the quality of qPCR reagents or mistakes in the setup of reactions should be considered.

Control of Inhibition

The described analysis setup also allows for the control of the potential presence of inhibitors in the sample that can influence the efficiency of qPCR. Inhibition is checked only in assays for taxon-specific sequences. In our case, one extraction parallel is analyzed in a stock solution (see the section, Homogenization and DNA Extraction) and tenfold dilution in duplicate (Table 10). The second extraction parallel is tested only in a stock solution, in duplicate. To check for possible inhibition, the ΔCq (sample) for the taxon-specific gene is calculated:

$$\Delta Cq(sample) = Cq \text{ (stock solution DNA)} - Cq \text{ (}10\times \text{ dilution)}$$

If different dilutions are used, the slope is calculated as follows.

$$\text{Slope} = SLOPE \text{ (Cq, log}_{10}\text{relative concentration)}$$

The *SLOPE* function returns the slope of the linear regression line through data points in known_*y*'s and known_*x*'s. The slope is the vertical distance divided by the horizontal distance between any two points on the line, which is the rate of change along the regression line.

Table 11 Data from *hmgA* analysis and calculated parameters. k: slope, R_1^2: R^2 coefficient, n: intercept from the Cq values and $\log_{10}$ copy number of positive DNA target control. 1×: stock solution, 10×: 10× dilution

hmgA					
Positive DNA target control RM	Cq	Copy number*	$\log_{10}$ copy number	Parameters	
ID184 1×	24.65	50000	4.69897	k_1	−3.32
ID184 1×	24.65	50000	4.69897	R_1^2	0.999999
ID184 10×	27.97	5000	3.69897	k_2	−0.301205
ID184 10×	27.97	5000	3.69897	n_2	12.1236

*Reference material ID 184 has known (estimated) *hmgA* copy number

If ΔCq or the slope is between −2.9 and −4.1 we conclude that no inhibition in the sample was detected.

In this first step of qPCR analysis, inhibition is checked primarily as valuable information for further quantitative testing. Qualitative analyses do not need to be repeated, except in the case of strong inhibition, such as if the Cq value of 10× dilution is lower than that of the stock solution. In this case, strong inhibition of qPCR reaction for the transgene is possible and might give a false negative result in screening analysis. The analysis should be repeated with more diluted DNA. If strong inhibition is still detected, DNA should be extracted using a different method. However, this was never the case in our routine work. Inhibition control is not needed for qualitative analysis of screening elements (P-35S, T-NOS), because results are not informative for further quantitative analysis.

Practical LOD and LOQ

In cases where the target transgene is not detected, additional information is required in the report. In addition to absolute LOD of the method (determined on reference material during verification of the method), practical LOD is calculated for each test sample. The practical LOD is the lowest relative quantity of the target DNA that can be detected, given a known (determined/estimated) number of target taxon genome copies. The practical LOQ is the lowest relative quantity of the target DNA that can be reliably quantified, given a known (determined/estimated) number of target taxon genome copies (International Organization for Standardization 2006). Namely, practical LOD/LOQ can differ from LOD/LOQ of the method because it is also related to the sample type and the quality/quantity of the template DNA (International Organization for Standardization 2006).

The practical LOD and LOQ are calculated on the data obtained from analysis of the taxon-specific sequence on positive DNA target control, where copy numbers are known. First, slope k_2 and intercept n_2 are calculated from the Cq values and log copy number of positive DNA target control (Table 11).

$$k_2 = SLOPE \ (\log_{10}\text{copy number; Cq value})$$

$$n_2 = INTERCEPT \ (\log_{10}\text{copy number; Cq value})$$

The *INTERCEPT* function calculates the point at which a line will intersect the *y*-axis by using existing *x*-values and *y*-values. The intercept point is based on a best-fit regression line plotted through the known *x*-values and known *y*-values.

Using slope k_2 and intercept n_2 the taxon-specific sequence copy numbers in the test sample are calculated and from these, the practical LOD/LOQ are calculated (Table 12).

$$hmgA \text{ copy number } = 10^{(\text{Cq value} * k_2 \ (hmgA) + n_2 \ (hmgA))}$$

Practical LOD = LOD of the method / *hmgA* average copy number per reaction*100

Practical LOQ = LOQ of the method / *hmgA* average copy number per reaction*100

As already mentioned, the absolute LOD and LOQ depend on the method used. They are determined during the verification of the method (see the section, Verification of Methods). In our laboratory, the absolute LOD for method for *hmgA* is two copies, and the absolute LOQ is 50 copies.

Negative results should not give an unjustified impression of the reliability (Holst-Jensen 2009). For this reason, we have decided to report the highest practical LOD/LOQ of the sample if possible. This means that the highest practical LOD/LOQ calculated is reported. In the case of inhibition, the value of the 10× diluted sample DNA divided by the dilution factor is used, as exemplified in the following. Data in Table 10 show that there is weak inhibition in undiluted samples, therefore we would take results from the 10× diluted sample and divide it by the dilution factor to evaluate the practical LOD/LOQ. Practical LOD and LOQ for the sample presented in Table 12 are 0.001% and 0.02%, respectively.

Qualitative Analysis: Event-Specific Methods

Where the presence of screening elements is confirmed in a test sample, presence of specific transgenic lines is tested. The decision on which event-specific methods should be performed is based on the combination of positive/negative results determined from the screening step. As an example, if we detect the presence of both screening elements (P-35S and T-NOS) in a maize flour sample, we test the 12 maize lines (Table 4) harboring P-35S, T-NOS, or both elements. If only P-35S were positive, we would test for five lines that contain only P-35S.

For each specific event, analysis is performed similarly and with the same criteria as for the screening elements. The qPCR performance is controlled with NTC for each analyte at the beginning and at the end of pipetting, and with positive target

Table 12 *HmgA* copy numbers and practical LOD and LOQ

hmgA				
Sample name	Copy number	Average copy number	Practical LOD (%)	Practical LOQ (%)
Sample1- parallel1 1×	600221	548804	0,0004	0,01
Sample1- parallel1 1×	497386			
Sample1- parallel1 10×	23402	24260	0,0082	0,21
Sample1- parallel1 10×	25117			
Sample1- parallel2 1×	399250	428715	0,0005	0,01
Sample1- parallel2 1×	458180			

DNA control for each analyte in duplicate. Inhibition is not tested again. However, if inhibition were observed during the screening step, more diluted DNA might be used for detection of specific events.

Semiquantitative Analysis

Over several years of routine analysis, we have noticed that in our laboratory samples typically fall into two groups regarding their GMO content: they usually have either very high (decisively above 0.9%) or very low GMO contents (below 0.1%). Based on this experience and in order to cost optimize the work in our laboratory, we have decided to introduce semiquantitative analysis for samples for which very low levels of GMOs are expected based on the screening step. We have investigated our records from quantitative analysis to see if a criterion could be set for semiquantification of samples based on the results of a qualitative analysis setup. Differences in Cq values between transgene and taxon-specific sequences were calculated by subtraction of the average transgene Cq value from the average taxon-specific sequence Cq value. Potential different efficiencies of transgene and taxon-specific sequence were also taken into account. For maize, when the taxon-specific sequence Cq value differs from the transgene Cq value for more than 13.0, it is possible to conclude without further analysis that the sample contains less than 0.1% GM maize. If the difference between Cq values is below 13.0, we continue analysis with quantification of specific events that were identified in qualitative analysis.

Each laboratory wishing to apply such a semiquantitative approach should use its own data inasmuch as our system depends on methods applied and the system used for analysis of raw qPCR data (e.g., setup, threshold, and baseline value) used in our laboratory.

Quantitative Analysis

During the quantification step, the ratio between the amount of transgene copies and the amount of taxon-specific sequence copies is determined. Quantitative analysis usually follows qualitative analysis which means that we already have collected some data about the quality of the DNA from test samples. However, quantification is the most demanding step in detection of GMOs. That is why additional quality controls are used.

In our laboratory, the standard curve approach (Applied Biosystems 2001) is used for quantification. We have chosen this approach over the $\Delta\Delta$ Cq approach. The $\Delta\Delta$ Cq method assumes that both assays have similar efficiencies of amplification. Therefore the approach would only be acceptable when working with well-established samples (e.g., with raw materials) and if the qPCR methods are validated in combination with the DNA extraction method (nonmodular approach). In the standard curve approach, the difference in amplification efficiency between the analytes is taken into account. How qPCR efficiency can be influenced is well described in Cankar et al. (2006) and Demeke and Jenkins (2009).

Sample Analysis

Returning to our maize flour sample, if we wish to confirm, for example, the presence of MON-00810-6 in the sample we would proceed with quantification. Quantitative analysis is performed in 20-μl reactions. Three dilutions of each extraction parallel are tested in duplicate. For samples that are analyzed frequently or known samples that usually do not contain inhibitors, two dilutions may be enough. However, from our experience, analysis in three dilutions often prevents repetition of analysis and is helpful in the final decision regarding the analytical result.

Quality Controls

In parallel with analysis of test samples, NTC and dilutions of reference materials are tested. NTC for each analyte is tested in duplicate, one at the start of pipetting and one at the end. NTC should result in no amplification, otherwise the analysis is not valid and contamination of the reaction mix should be considered.

The standard curve is prepared using CRM. Five dilutions of CRM DNA are prepared (Table 13) in such a way that the first dilution contains approximately 100 ng of DNA per reaction (stock solution) and that all measurements are in the linear range of the method. As an example, from our experience we know that for 5% MON-00810-6 maize genomic DNA, 3×, 9×, 27×, and 81× dilutions must be prepared. All five dilutions are analyzed in duplicate for the taxon-specific sequence and transgene. It is important to be aware that the above-mentioned dilutions are fit for our system and for our most frequently used method of extraction. Appropriate dilutions depend on extraction efficiency, therefore each laboratory should set its own system of dilutions to best cover the linear range of the method.

An appropriate value is assigned to each dilution in accordance with the GMO content in CRM (5% in this case). The stock solution has value 100 for *hmgA* and 5 for MON-00810-6. Three times lower values are assigned to 3× dilutions. The logarithm of value is calculated (Table 14).

Table 13 An example of qPCR standard curve result. Dilutions of MON-00810-6 CRM and their Cq. 1×: stock solution, 3×, 9×, 27×, 81×: 3× serial dilutions

	hmgA	MON-00810-6
Sample name	Cq	Cq
MON-00810-6 1×	21.21	26.98
MON-00810-6 1×	21.20	27.13
MON-00810-6 3×	23.02	28.33
MON-00810-6 3×	22.98	28.36
MON-00810-6 9×	24.75	29.99
MON-00810-6 9×	24.58	30.06
MON-00810-6 27×	26.08	31.29
MON-00810-6 27×	26.17	31.54
MON-00810-6 81×	27.73	32.93
MON-00810-6 81×	27.72	33.30

Table 14 Values assigned to dilutions of CRM and their logarithms (base 10) calculated. 1×: stock solution, 3×, 9×, 27×, 81×: 3× serial dilutions

	hmgA			MON-00810-6		
Sample name	Cq	Value	Log_{10} value	Cq	Value	Log_{10} value
MON-00810-6 1×	21.21	100	2	26.98	5	0.699
MON-00810-6 1×	21.20	100	2	27.13	5	0.699
MON-00810-6 3×	23.02	33.33	1.523	28.33	1.667	0.222
MON-00810-6 3×	22.98	33.33	1.523	28.36	1.667	0.222
MON-00810-6 9×	24.75	11.11	1.046	29.99	0.556	−0.255
MON-00810-6 9×	24.58	11.11	1.046	30.06	0.556	−0.255
MON-00810-6 27×	26.08	3.704	0.569	31.29	0.185	−0.732
MON-00810-6 27×	26.17	3.704	0.569	31.54	0.185	−0.732
MON-00810-6 81×	27.73	1.235	0.092	32.93	0.062	−1.209
MON-00810-6 81×	27.72	1.235	0.092	33.30	0.062	−1.209

From these data for the CRM, slope k_1 and correlation coefficient R_1^2 of the standard curves are calculated for the taxon-specific sequence (*hmgA*) and the transgene (MON-00810-6), respectively.

$$k_1 = SLOPE(\text{Cq}; log_{10}\text{value})$$

$$R_1^2 = RSQ(\text{Cq}; log_{10}\text{value})$$

The *RSQ* function returns the square of the Pearson product moment correlation coefficient through data points in known_*y*'s and known_*x*'s. The *r*-squared value can be interpreted as the proportion of the variance in *y* attributable to the variance in *x*.

The standard curve slope k_1 must be between −3.0 and −3.8, and the correlation coefficient $R_1^2 > 0.98$ (Fig. 15), otherwise the quantification needs to be repeated. The same criteria apply for both the taxon-specific sequence and the transgene.

Fig. 15 Standard curve. Slope k_1 and correlation coefficient R_1^2 of the standard curve are calculated from data on Cqs and logarithm (base 10) of assigned values

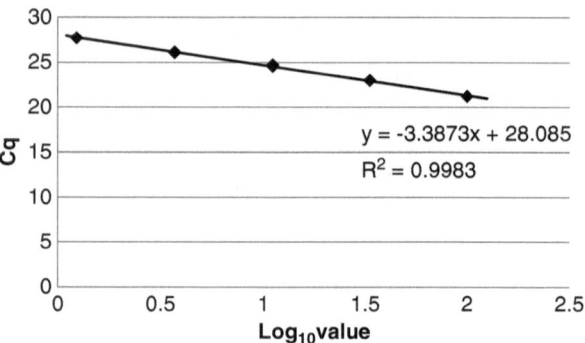

Then, the slope k_2 and the intercept on the ordinate axis n_2 for each standard curve are calculated.

$$k_2 = SLOPE \ (\log_{10} \text{value; Cq})$$

$$n_2 = INTERCEPT \ (\log_{10} \text{value; Cq})$$

These values are used for the calculation of the copy numbers of the taxon-specific and transgene sequences for standard curves (Table 15) and later on for samples.

$$\text{taxon - specific sequence copy number} = 10^{(Cq*k_2 \ (hmgA)+n_2 (hmgA))}$$

$$\text{transgene copy number} = 10^{(Cq*k_2 \ (MON-00810-6)+n_2 \ (MON-00810-6))}$$

The average and standard deviation of the taxon-specific sequence and transgene copy number are calculated for each point of the standard curve and later also for samples to determine the coefficient of variation (CV).

$$CV = \text{(standard deviation)} / \text{(average copy number of the target sequence)}$$

If the CV of the taxon-specific sequence or transgene copy number of the outer point of the standard curve is above 0.25, then this point is outside the linear range of the method and must be excluded from further calculations. If the CV of the average copy number of the taxon-specific sequence or transgene of the internal point of the standard curve is above 0.25 and that one value for this point is an obvious outlier (e.g., pipetting error), this value is excluded from the standard curve. If no obvious outlier is identified, both replicate values for this point should be excluded from the calculation of the standard curve.

Sample Analysis

If all quality controls are in accordance with the acceptance criteria, the data from samples are checked and the ratio between the transgene and taxon-specific

Table 15 Taxon-specific sequence and transgene copy number calculations. 1×: stock solution, 3×, 9×, 27×, 81×: 3× serial dilutions

Sample name	hmgA				MON-00810-6			
	Cq	Value	Log_{10} value	Taxon-specific sequence copy number	Cq	Value	Log_{10} value	Transgene copy number
MON-00810-6 1×	21.21	100	2	106.57	26.98	5	0.699	4.84
MON-00810-6 1×	21.20	100	2	107.42	27.13	5	0.699	4.36
MON-00810-6 3×	23.02	33.33	1.523	31.30	28.33	1.667	0.222	1.84
MON-00810-6 3×	22.98	33.33	1.523	32.06	28.36	1.667	0.222	1.79
MON-00810-6 9×	24.75	11.11	1.046	9.68	29.99	0.556	−0.255	0.56
MON-00810-6 9×	24.58	11.11	1.046	10.85	30.06	0.556	−0.255	0.53
MON-00810-6 27×	26.08	3.704	0.569	3.92	31.29	0.185	−0.732	0.22
MON-00810-6 27×	26.17	3.704	0.569	3.68	31.54	0.185	−0.732	0.18
MON-00810-6 81×	27.73	1.235	0.092	1.28	32.93	0.062	−1.209	0.07
MON-00810-6 81×	27.72	1.235	0.092	1.29	33.30	0.062	−1.209	0.05

sequence copy number is calculated (Fig. 16). The taxon-specific sequence and transgene Cq values for each sample are first checked. For quantification, only the Cq values within the linear range can be used. Then, efficiencies of amplification of each sample are checked. As was mentioned above, there might be differences in amplification efficiencies due to coextracted impurities that affect the kinetics of the amplification process. Therefore, we calculate slopes (k_1) from the dilutions of test samples for both the transgene and the taxon-specific sequence.

We can then proceed with quantitative analysis when

$$| k_1 (\text{sample}) - k_1 (\text{standard curve})| \leq 0.3$$

If the difference between the slope for the test sample and the one for the standard curve is above 0.3, we consider that the GMO content cannot be quantified. The analysis should be repeated.

If there is no inhibition, we can proceed with calculations. Quantities of taxon-specific sequence copies and transgene copies are calculated using slope (k_2) and the intercept (n_2) from the taxon-specific sequence and transgene standard curves, respectively, as previously described. From the taxon-specific sequence or transgene copy numbers, the average and standard deviation are calculated and the CV is determined. Only the results of amplification with CV for the taxon-specific

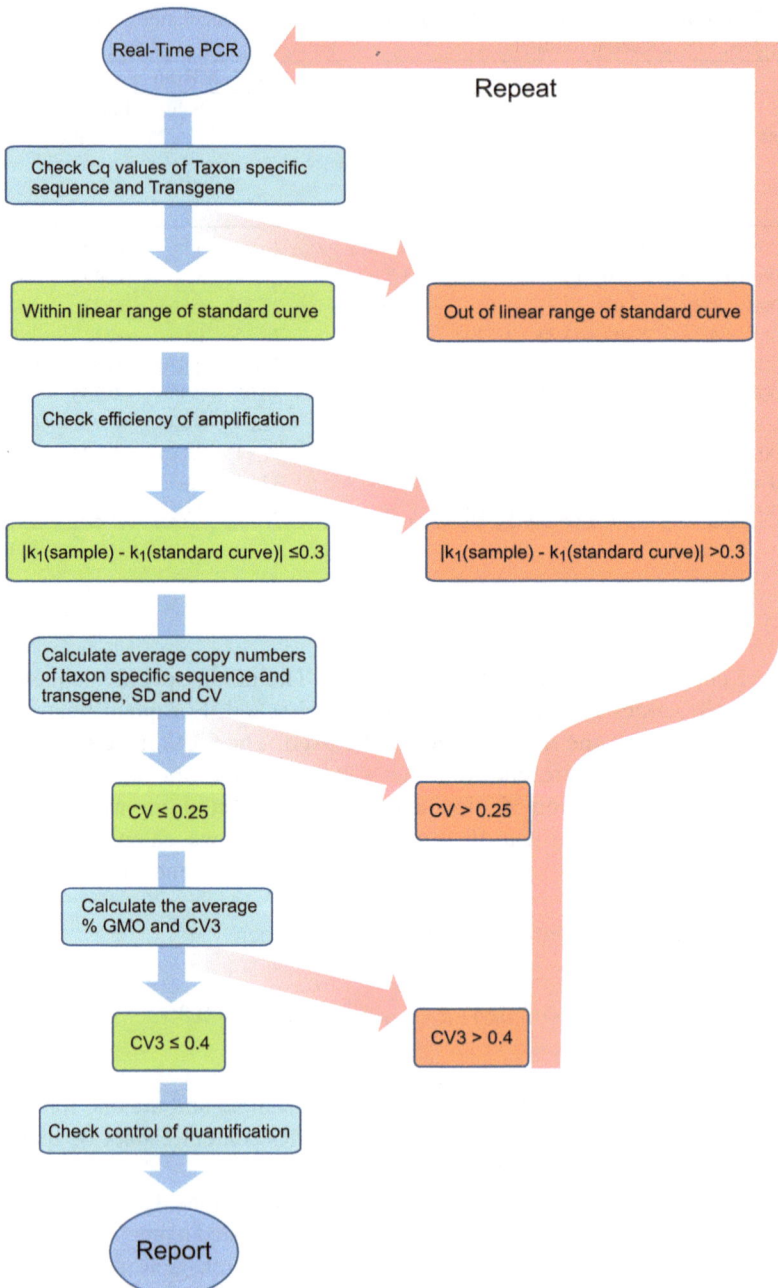

Fig. 16 Flowchart and decision support for sample analysis using qPCR

sequence or transgene copy number below 0.25 can be used for quantification. Finally, transgene content is calculated for both parallels and for all dilutions:

$$\%GMO = (\text{average copy number of } hmgA)/$$
$$(\text{average copy number of MON-00810-6}) * 100$$

If the dilutions used for the transgene and taxon-specific sequence copy number determination are different, the difference in dilution factor has to be taken into account for the final GMO content calculation. In addition, the homogeneity of the analytical sample should be controlled. For this purpose the coefficient of variation between the transgene content of both extraction parallels (CV3) is calculated.

$$CV3 = STDEV(\text{GMO\% 1st parallel; GMO\% 2nd parallel})/\text{average GMO\%}$$

The STDEV function estimates the standard deviation based on a sample. The standard deviation is a measure of how widely values are dispersed from the average value (the mean).

If CV3 is above 0.40 the analytical sample is considered as heterogeneous. In that case, the analytical sample should be mixed again, new test portions for DNA extraction taken, and the whole analysis repeated.

There are many factors influencing each measurement. Therefore the final result of testing should include data on MU. See the section, Measurement Uncertainty.

Control of Quantitative Performance of the Analysis

A specific quality control used during this analytical step is the control of quantification. This is the sample with known transgene content which means it is a quantification control sample (see Organization of the Laboratory and Quality Management System). Two dilutions of the quantification control sample are tested in duplicate. Inhibition is checked and the ratio between transgene and taxon-specific sequence is calculated as done for test samples. The result of the quantification control sample should not differ more than 50% from the assigned (known) value. If this requirement is not fulfilled, calculations should be checked again and if no calculation error explains the difference, the quantitative analysis should be repeated.

Reporting

A report should, in addition to the analytical result, also include data on uncertainty and error associated with a measurement or data on reliability of the negative result (Holst-Jensen 2009). Reports are written in accordance with the following standards: EN ISO/IEC 17025 (International Organization for Standardization 2005d), EN ISO/IEC 24276 (International Organization for Standardization 2006), EN ISO/IEC 21569 (International Organization for Standardization 2005a), and EN ISO/IEC 21570 (International Organization for Standardization 2005a). Reports should

Table 16 Example of information on sample identity

Report No.: GMO001/11	
Customer:	Inspection service Ljubljana
Contact person:	Mr. NOVAK
Sample number:	G042/10
Customer's designation of the sample:	045768
Date of sample receipt:	09.03.2011
Sample specifications:	Maize kernels
Sample quantity:	1 kg
Analysis order:	P-35S, T-NOS screening analysis
	Determination of specific transgene lines
	Indirect quantitative analysis
Date of analysis:	11.3.2011–15.03.2011

Table 17 Example of information on the methods used for analyses

Test methods performed in accordance with standard operation procedures	
02G-Pos06	Extraction and purification of DNA by CTAB method
02G-Pos53	Quantitative determination of maize reference gene (*hmgA*) by qPCR
02G-Pos47	Qualitative determination of P-35 S by qPCR
02G-Pos15	Qualitative determination of T-NOS by qPCR
02G-Pos19	Quantitative determination of maize MON-00810-6 by qPCR

Table 18 Example for expression of negative result

Screening analysis		
Amplicon	Result	Limit of Detection (LOD) of the method
T-NOS	Not detected	2 copies of target DNA
P-35S	Not detected	2 copies of target DNA

Limit of detection of these methods was determined on certified reference material. Practical limit of detection for maize lines is 0.002%

Table 19 Example of expression of result for quantitative analyses

Quantitative analysis					
Amplicon	Result	%GMO	Expanded measurement uncertainty	Limit of Detection (LOD) of the method	Limit of Quantification (LOQ) of the method
MON-008 10-6	Detected	7.7	±2.3	2 copies of target DNA	50 copies of target DNA

Limit of detection and limit of quantification of the method were determined on certified reference material. Expanded measurement uncertainty is calculated on the basis of analyzed samples from previous years (coverage factor is 2, level of confidence is 95%)

include information needed to identify the laboratory sample (Table 16), and the methods used for the extraction of DNA and detection of GMOs (Table 17). If GMO was not detected, the LOD of the method obtained during verification of method and the practical LOD for the sample should be reported (Table 18). In the case of quantification of GMO, measurement uncertainty should be stated (Table 19).

Measurement Uncertainty

Introduction

Repeated measurements of the GMO content of a given sample, in the same or different laboratories, are not expected to produce a single GMO content value. Dispersion of the analytical values is usually observed. The measurement uncertainty (MU; also mentioned as uncertainty of measurement or simply uncertainty) is a nonnegative parameter characterizing the dispersion of the quantity values being attributed to a (International Organization for Standardization 2007; Fig. 17).

The sources influencing MU come from individual factors during the whole analytical procedure. They can be seen as pieces of a puzzle forming the final picture. Each analytical step, such as DNA extraction or qPCR, contributes to the final MU. In each analytical step, many individual factors influence the MU. Some examples of these influencing factors are listed here: laboratory sample homogeneity/heterogeneity, repeatability of DNA extraction, repeatability of different qPCR apparatus, chemicals, operators, and MU of CRMs. A large contribution to the MU is also due to the type of material sampled, especially where unequal distribution of GMO in non-GMO material is present. However, this book only focuses on the laboratory part of GMO detection and therefore, sampling MU is not covered.

Another important parameter influencing the reliability of the result is bias. Bias is defined as the difference between the test result or measurement result and a true value (International Organization for Standardization 2011). Bias can be seen as a deviation of a result from the target value (Fig. 18).

The laboratories shall have the procedure to estimate the MU in place (Codex Alimentarius Commission 2004; International Organization for Standardization

small MU   large MU

Fig. 17 Results of repeated measurements of the same sample can give very distributed values around target value, representing large MU, whereas results concentrated near the target value represent small MU

Fig. 18 Presentation of bias.
All the measurement results
are concentrated around one
virtual point that is not the
target concentration. The
difference between this
virtual point and the real
target value is bias

2005d). The MU is not an absolute value, but is always an estimation of value and is obtained from available data. It also has to be re-evaluated periodically based on newly obtained data.

MU in GMO Detection

Different approaches to calculate MU exist (EURACHEM-CITAC Working Group 2000; Priel 2009). One of them is the so-called bottom-up approach that calculates all the individual factors influencing the final result. A much more appropriate and practical approach to GMO detection is the so-called top-down approach. In this approach, data from collaborative trials and sample analyses, including all the factors influencing the MU during the analytical procedure, are used as a source for estimation of MU. This approach is described in detail in the guidance document on measurement uncertainty for GMO testing laboratories produced by the ENGL working group on measurement uncertainty (Trapmann et al. 2009).

Using Data from Collaborative Trials

In the top-down approach, the method performance data obtained from collaborative trials, typically data on repeatability, reproducibility, and trueness of the method can generally be used as valuable sources for MU estimation (International Organization for Standardization 2010). In order to use this MU calculation approach, the laboratory must ensure that the performance of the method implemented in the laboratory is consistent with the performance of the method measured during the collaborative trial. This is evaluated through measurement of the method bias and precision in the laboratory (International Organization for Standardization 2010). In addition, any factor that could potentially influence the measurement results and that was not adequately covered by the collaborative study should be identified. The variance of the results associated with this/these factor(s) should be quantified (International Organization for Standardization 2010).

The EU-RL GMFF systematically organizes collaborative trials of GMO detection methods and publishes the validation data on its Internet page (http://gmo-crl.jrc.ec. europa.eu/statusofdoss.htm; also see Verification of Methods). The ENGL and the EU-RL GMFF have established the method acceptance criteria and the method performance requirements for quantitative event-specific GMO detection methods. The method performance criteria in these trials are the following. The relative reproducibility standard deviation (RSD_R) should be <35% over the whole dynamic range (at concentrations <0.2%, RSD_R <50% are deemed acceptable), and trueness shall be within ±25% of the accepted reference value over the whole dynamic range (CRL-GMFF 2008). Bias was rarely observed and reported in these collaborative trials. Additional data from other collaborative trials are available (International Organization for Standardization 2005b; Kodama et al. 2010). Therefore, the data obtained from these collaborative trials can be used to calculate MU according to the top-down approach.

At the time the ENGL guidance document on measurement uncertainty for GMO testing laboratories was published, there were not enough data obtained from collaborative trials to calculate a common value that could be used as an estimate of the MU in GMO testing based on collaborative trials (Trapmann et al. 2009).

Recently Kodama and collaborators collected data from 53 collaborative studies published in the *Journal of AOAC INTERNATIONAL*, authorized by the International Organization for Standardization, and/or performed by the EU-RL GMFF (Kodama et al. 2010). These collaborative studies were performed on seven crops (maize, soya bean, cotton, oilseed rape, potato, sugar beet, and rice). The GMO content of the tested samples ranged from 0.02% to 10%. Some of the collaborative trials included both DNA extraction and PCR modules; others only included PCR on already extracted DNA solutions. The authors of this comparison concluded that the reproducibility standard deviation (S_R) and repeatability standard deviation (S_r) of the GMO amount are more or less independent of the plant species. The extraction step does not significantly increase S_R values, the PCR step being the main source of variation between measurement values. S_R and S_r are functions only of the GMO content. The authors found that RSD_R in each interlaboratory study was reduced for most GM events when the analyzed samples contained higher GMO contents. From all the analyzed data they proposed

$$S_R = 0.1971C^{0.8685} \quad and \quad RSD_R = 19.71C^{-0.1315}$$

where C is the GMO content (in%).

Applicability range of the proposed formulas is 0.1–10% of GMO content. Using this approach, Kodama and collaborators have estimated that the RSD_R at the level of 1% GMO is 20%, and at the level of 5% GMO is 16%.

The study by Kodama and collaborators provides a good basis for obtaining more harmonized estimates of MU in GMO testing, although the authors stated that further experimental confirmation of their proposal is needed because GMO analysis is a relatively new field, and the number of collaborative studies for GMO analysis will increase in the future (Kodama et al. 2010).

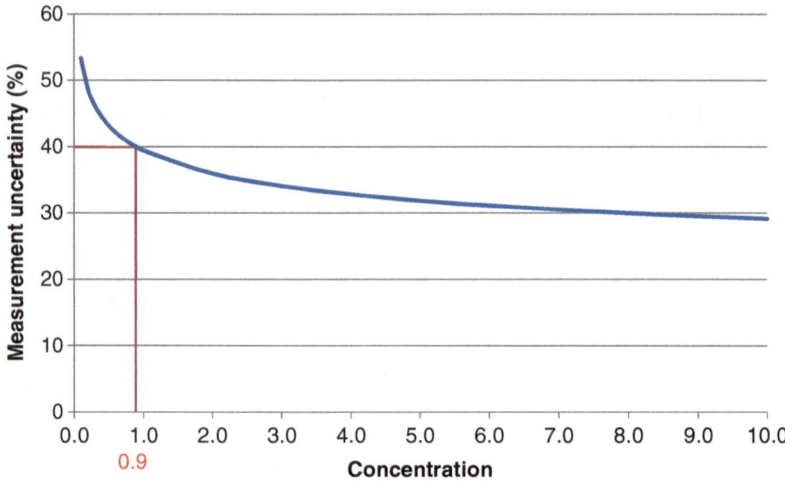

Fig. 19 The relative MU at different concentrations of GMOs. Emphasized relative MU at 0.9% GMO is 40%

Using the data from Kodama and collaborators, assuming the absence of bias and that the absolute uncertainty u_0 is negligible, the standard uncertainty u can be approximated by S_R as described in equation 12 of the ENGL guidance document on measurement uncertainty for GMO testing laboratories (Trapmann et al. 2009).

The expanded measurement uncertainty (U) at 95% confidence level, is obtained by multiplying the standard uncertainty u by a coverage factor $k=2$.

$$U = 2 * S_R = 0.3942 C^{0.8685}$$

The relative MU, expressed as U/c is shown in Fig. 19. In this example it can be seen that relative MU at 0.9% GMO is 40%, and above 8% GMO relative MU is around 30%. Below 0.9% GMO, relative MU is increasing, reaching 53% at 0.1% GMO.

Calculation of MU from Internal Quality Control Data

Before using the MU data obtained from collaborative trials, one needs to estimate the MU of analyses performed in its own laboratory. The estimation procedure described here is based on the handbook for calculation of MU in environmental laboratories (Magnusson et al. 2004), further adapted for GMO testing (Trapmann et al. 2009; Žel et al. 2007).

To estimate the MU of analyses performed in the laboratory, the data obtained during method verification and data from sample analyses can be used for internal quality control data. When a new method is implemented in the laboratory, only the data obtained during method verification are available. These data can afterwards be complemented with results of the analyses after initializing this method on laboratory samples. Periodically, new data from new analyses are added and the MU estimation is verified or corrected. Data collected from quantification control samples (see the section, Organization of the Laboratory and Quality Management System) allow estimation of the MU for these control samples. This MU from quantification of control samples can be compared with MU obtained from samples used during analyses.

An example of calculation of MU using data from routine samples (modified from Žel et al. 2009 and Trapmann et al. 2009) is presented below. In this example bias was not observed.

Two measurement results of GMO content in each sample are obtained (c_1 and c_2; Table 20). The mean (*Mean c*) from c_1 and c_2 for each sample is calculated as

$$Mean\, c = (c_1 + c_2)/2$$

If we take data from Table 20 as an example of how to calculate MU in practice, for example,

$$Mean\, c = (0.155 + 0.146)/2 = 0.150$$

The absolute difference (*d*) between c_1 and c_2 is calculated:

$$d = |c_1 - c_2| \qquad d = |0.155 - 0.146| = 0.009$$

The relative difference between c_1 and c_2 is calculated as

$$rad = (d\,/\,Mean\,c)*100 \qquad rad = (0.009\,/\,0.150)*100 = 5.707$$

From a set of calculated differences (*d*) and relative differences (*rad*), the average difference (*Mean d*) and average relative difference (*Mean rad*) are calculated (for calculation of the data in Table 20, see there).

The within laboratory standard deviation (s_R) is calculated from *Mean d* divided by factor d_2 (1.13 in the case of two independent measurements):

$$s_R = Mean\, d\,/\,d_2 \qquad s_R = 2.731/1.13 = 2.417\%$$

Within laboratory reproducibility relative standard deviation (RSD_R) is calculated using *Mean rad* and d_2:

$$RSD_R = Mean\, rad\,/\,d_2 \qquad RSD_R = 14.140/1.13 = 12.513\%$$

Table 20 Measurement results obtained on routine samples ($n=2$) and calculation of the relative difference

Analysis number	GMO concentration c_1	GMO concentration c_2	Mean GMO concentration Mean c [%]	Differenced [%]	Relative difference rad [%]
G018/05	0.155	0.146	0.150	0.009	5.707
G199/06	0.143	0.178	0.161	0.035	21.737
G128/09	0.200	0.210	0.205	0.010	4.878
G098/05	0.282	0.339	0.311	0.057	18.381
G142/05	0.344	0.285	0.314	0.059	18.613
G126/08	0.415	0.292	0.354	0.123	34.795
G162/08	0.456	0.556	0.506	0.100	19.763
G101/05	0.440	0.638	0.539	0.198	36.729
G252/06	0.738	0.651	0.695	0.087	12.468
G121/05	0.669	0.890	0.780	0.221	28.282
G230/06	1.082	1.490	1.286	0.407	31.680
G133/06	1.262	1.511	1.386	0.249	17.976
G013/05	1.948	1.250	1.599	0.698	43.681
G115/05	1.582	1.677	1.630	0.094	5.793
G237/06	2.117	2.141	2.129	0.024	1.126
G175/05	2.174	2.226	2.200	0.052	2.356
G176/05	2.371	2.157	2.264	0.215	9.492
G015/05	2.322	2.456	2.389	0.134	5.606
G100/05	3.332	1.922	2.627	1.410	53.682
G247/06	2.891	2.548	2.719	0.344	12.639
G104/05	4.321	3.856	4.088	0.465	11.362
G103/04	4.901	4.776	4.839	0.125	2.581
G117/05	5.234	5.464	5.349	0.230	4.300
G118/06	12.514	14.418	13.466	1.904	14.143
G116/05	17.462	17.889	17.675	0.427	2.416
G119/05	19.902	19.589	19.745	0.313	1.586
G021/05	23.143	20.127	21.635	3.016	13.942
G159/08	21.675	22.608	22.142	0.933	4.214
G099/05	24.448	22.302	23.375	2.146	9.179
G102/05	36.473	41.474	38.973	5.001	12.832
G072/08	46.502	44.409	45.456	2.093	4.605
G124/08	48.186	49.881	49.034	1.695	3.457
G131/08	51.450	57.181	54.316	5.731	10.551
G133/05	63.665	54.036	58.850	9.629	16.362
G101/08	61.346	61.294	61.320	0.052	0.085
G135/05	56.809	71.972	64.391	15.163	23.549
G112/05	65.804	66.282	66.043	0.477	0.723
G174/08	68.788	75.231	72.010	6.443	8.947
G093/08	86.590	85.354	85.972	1.236	1.438
G158/08	114.636	81.385	98.011	33.251	33.926
Mean			**21.273**	**2.371**	**14.140**

The expanded uncertainty U, corresponding to a confidence level of approximately 95%, is obtained by multiplication of RSD_R by coverage factor $k = 2$.

$$U = RSD_R * 2 \qquad U = 12.513 * 2 = 25.026\%$$

At present in our laboratory, estimated expanded measurement uncertainty is 30% for a GMO content above 0.4% and it is 45% for a GMO content below 0.4%. Data on MU estimated from laboratory samples and the ones estimated from the quantification control sample are in the same range. It can be expected that MU in one laboratory is lower than one estimated from collaborative trials, where more laboratories are performing analyses with differently experienced personnel, using different equipment, chemicals, and so on. This is the case for our laboratory in comparison with the MU estimation from Kodama's (2010) paper.

Reporting of MU

When reporting MU in the final test report to the customer, it is important to explicitly provide the source of data used for the calculation of MU (e.g., "MU was calculated on the basis of collaborative trials," or "MU was calculated on the basis of internal quality control data"). Reporting the data source used for the estimation of MU is important as it allows comparisons of the results obtained in different laboratories. The coverage factor used for the estimation of the extended MU should also be stated (e.g., "a coverage factor of 2 was used, corresponding to a confidence level of approximately 95%"). An example can be seen in the section, Real-Time PCR, Table 19.

MU and Compliance of Results with Legislation

To decide whether the GMO content in a sample is in compliance with legislation requirements, it is necessary that the analytical result take into account the measurement uncertainty (EURACHEM-CITAC Working Group 2007). In the EU legislation regarding GMO content in food and feedstuff, it is not specified if the MU needs to be subtracted or added to the reported concentration. In the ENGL guidance document on MU for GMO testing laboratories, it is proposed that the GMO content value obtained by subtracting the expanded uncertainty from the measured GMO concentration is used to assess compliance with the EU legislation (Trapmann et al. 2009). Only when this value is greater than the legal threshold, is it considered as sure "beyond a reasonable doubt" that the analyte content in the sample is beyond what is permissible (Fig. 20).

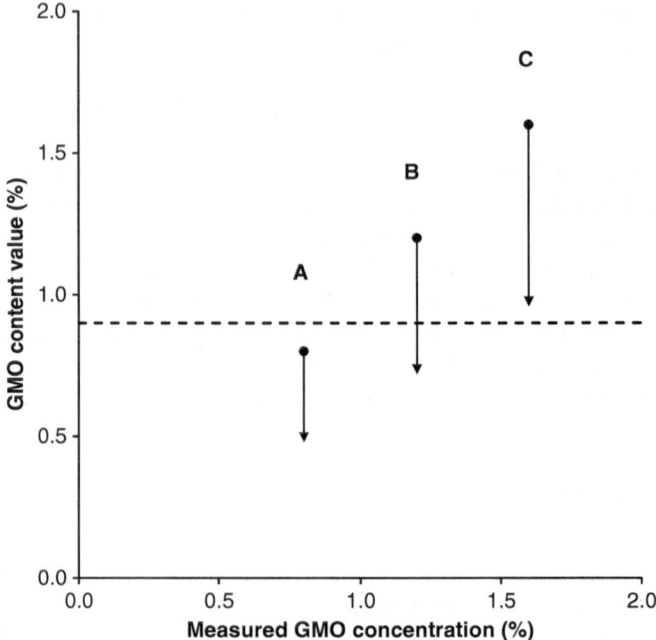

Fig. 20 Interpretation of GMO testing result for labeling compliance within EU legislation. GMO content value (*arrowhead*) obtained by subtracting the expanded uncertainty (*vertical line*, calculated using 40% as relative MU) from the measured GMO concentration (*circle*) is used to assess compliance with the EU legislation (*dashed line*). Samples with GMO content value (*arrowhead*) above the threshold value (*dashed line*) should be labeled, in this case only sample C

New Challenges

Introduction

Many challenges related to GMO analysis are being addressed, such as the detection of unauthorized GMOs, but new challenges are facing us. Each year, new GMOs are entering the world market. Their number and diversity are growing rapidly, and they challenge the established system of traceability and detection schemes. Apart from plants, other types of GMOs are on the market or close to commercialization, such as salmon and pig (Marris 2010). Most of the nonplant organisms that are genetically modified are engineered for purposes other than human nutrition, such as the fluorescent zebra fish grown for pleasure (http://www.glofish.com/) or the pigs contributing to reduction of phosphorus pollution (Golovan et al. 2001). Also, some plants and animals are used for "molecular pharming" to produce pharmaceutical proteins and chemicals (Faye and Gomord 2010; Sourrouille et al. 2009). New techniques for the development of plants and other organisms are already in research laboratories (Breyer et al. 2009; Rommens et al. 2007; Shukla et al. 2009), and the first products should be close to commercialization in the coming years.

These new GMOs present more diverse characteristics than the currently commercialized ones, challenging detection of these new GMOs. Further improvement of existing and development of new approaches for detection will be needed to face these upcoming situations.

In this section we emphasize some promising developments of methodologies and approaches that can be used in the near future to support detection of GMOs and we direct the reader to further recent sources of information.

Multiplex qPCR Methods

One of the possibilities for improving the efficiency of GMO diagnostics is to analyze several targets simultaneously. The ability to multiplex greatly expands the power of qPCR analysis, but requires evidence demonstrating that accurate amplification of multiple targets in a single tube is not impaired; that is, amplification efficiency and the sensitivity are the same as when the assays are performed in a singleplex qPCR setup (Bustin et al. 2009). This concern is of particular importance when targets of lower abundance are coamplified with highly abundant targets. Therefore, validation of multiplex methods needs additional parameters in order to be tested in comparison with singleplex methods.

A guidance document for the in-house and collaborative validations of multiplex qPCR methods was recently prepared by the members of the project GMOseek, under the European ERA-NET consortium SAFEFOODERA. Multiplex qPCR method performance parameters, the relevant verification procedures, and acceptance criteria were defined. This document should soon be made available to the public via the U.K. Food Standards Agency and the German Federal Office of Consumer Protection and Food Safety. Furthermore, an ENGL working group is currently re-evaluating the document "Definition of Minimum Performance Requirements for Analytical Methods of GMO Testing," taking into account the needs for implementation and validation of qualitative and multiplex qPCR methods for GMO detection.

Another obstacle for multiplexing also comes from the intrinsic technical limitations of qPCR machines. Currently, it is not possible to follow in real-time the amplification of more than five target analytes in a single reaction. This is due to the limited number of channels in the machine and the number of dyes available that can be clearly distinguished from each other. Fortunately, a new generation of qPCR platforms extends the optical range in which dyes are excited and fluorescence is detected from UV to infrared wavelengths, thus alleviating this obstacle (Debode et al. 2010).

In spite of the above-cited obstacles, multiplex qPCR methods for GMO detection have been developed allowing the simultaneous detection of two (Hohne et al. 2002; Waiblinger et al. 2008b; Zhang et al. 2008b), four (Dorries et al. 2010), or even six target analytes (Bahrdt et al. 2010). For example, the latter is a screening hexaplex assay covering more than 100 GMOs approved somewhere in the world. It comprises the following targets: P- 35S, T-NOS, Figwort Mosaic Virus 34 S promoter, and two construct-specific sequences present in novel GM soya bean and maize events that lack common screening elements. In addition, a detection system for an internal positive control (IPC) indicating the presence or absence of PCR-inhibiting substances is included (Bahrdt et al. 2010). However, it is important to note that so far, only one of these methods was fully validated in a collaborative trial (Waiblinger et al. 2008b) and, as explained above, current efforts to propose harmonized ways of fully validating such multiplex qPCR methods should stimulate their adoption in routine detection of GMOs.

Alternative PCR-Based Technologies

The Luminex xMAP technology combines advanced fluidics, optics, and digital signal processing with proprietary microsphere technology to deliver multiplexed assay capabilities. A screening protocol was developed that allows the simultaneous detection of specific P-35 S, and *epsps* DNA sequences in GM soya bean flours (Fantozzi et al. 2008). The entire procedure was composed of four fundamental steps: first, DNA targets were labeled by PCR; second, the selected sets of beads were coupled with probes; third, the hybridization step between labeled targets and coupled beads was performed; and finally, the detection step was performed using the Luminex-100 instrument. Even though the approach was only tested on a limited number of GM events, theoretically any additional primer/probe sets can be combined and the system can be easily adapted according to needs. The adoption of this approach however, although elegant and promising, is still hampered by the limited use of the Luminex device in GMO control laboratories (Querci et al. 2010).

The combination of PCR-based amplification and detection of amplicons with capillary gel electrophoresis can also be used for multitarget detection. As an example, the simultaneous detection of six cotton and five maize targets by multiplex PCR-capillary gel electrophoresis with identification of the amplified targets by size and color was reported (Nadal et al. 2006, 2009). More recently, the level of multiplexing using this approach was raised to 24 simultaneously detected DNA targets (Guo et al. 2011). This approach can be considered as a promising tool for GMO screening thanks to its flexibility: different multiplex PCR products can be combined in a single capillary gel electrophoresis with identification of amplified targets by size and color run. However, as for the Luminex-based technology, wide use of this approach may be hampered by the low availability of capillary electrophoresis devices in the laboratories. Also, both Luminex and capillary electrophoresis-based approaches allow only detection and not quantification of GMOs.

Ready-To-Use Multitarget Analytical System

The prespotted plates developed by the JRC are a qPCR-based ready-to-use multitarget analytical system for GMO detection (Querci et al. 2009). It allows simultaneous event-specific detection of 39 single-insert GMOs and their derived stacked events.

The system performance (specificity, efficiency, and LOD) has been successfully confirmed by experimental testing conducted within the EU-RL GMFF and in collaboration with European control laboratories. The system is based on TaqMan qPCR technology and consists of 96-well prespotted plates containing lyophilized primers and probes for the individual GMO detection. The specificity of each method (including the event-specific methods for 39 GM events, the P-35 S::bar construct-specific method, the taxon-specific methods for maize, cotton, rice, oilseed rape, soya bean, sugar beet, and potato) was confirmed. The LOD has been determined to be at least 0.045% expressed in haploid genome copies. The system presents a very promising approach for a harmonized detection system. A limitation could be the inflexibility of the current system, which is in further development. The system is based on 96-well plates, therefore a large amount of DNA is required to test all events spotted.

Microarrays

DNA microarrays (DNA chips) are high-throughput platforms enabling simultaneous detection of numerous targets on a single "chip." For GMO detection, the system usually consists of the preamplification step of the genetic targets, followed by hybridization on a "chip" containing specific oligonucleotide capture probes, and finally a detection step. There are several publications relating diverse GMO detection systems on microarray platforms as recently summarized by two reviews (Querci et al. 2010; von Gotz 2010). For one of these systems, a commercially available chip for GMO detection (DualChip® GMO, Eppendorf, Germany) combined with several multiplex PCR, an interlaboratory validation was successfully performed in 2008 (Leimanis et al. 2008), but wider usage of microarrays in GMO testing laboratories for routine analyses was not observed. So far, the use of microarrays for GMO detection offers only qualitative information (sometimes semiquantitative) and further improvement to quantification on a chip would be of high value. In addition to the usual obstacles linked to multiplexing amplification, the exponential nature of the PCR amplification limits the quantification potential of hybridized endpoint PCR products on microarrays. To circumvent this limitation, a novel multiplex quantitative DNA-based target linear amplification method, named NASBA implemented microarray analysis (NAIMA) was shown to be suitable for sensitive, specific, and quantitative detection of GMOs on microarrays (Dobnik et al. 2010; Morisset et al. 2008). This method shows potential for further development of quantification on microarrays but the system, so far limited to triplex amplification, should show a higher level of multiplexing to reach routine GMO detection laboratories.

In every case, the use of microarray-based technology for routine GMO diagnostics is restrained by the need for special and usually costly equipment for scanning microarrays. Another drawback is the possibility of cross-contamination during opening of the PCR tubes after the amplification step. The approach is also laborious in comparison with the cost and capacity of the system.

Digital PCR

Digital PCR (dPCR) is a relatively new technology, using PCR combined with partitioning of a single sample in a very large number of small chambers resulting in individual positive or negative amplification results in each chamber. It was shown that dPCR can be used for sensitive GMO detection (Bhat et al. 2009) and quantification (Corbisier et al. 2010) as well as assessment of detection limits in GMO analysis (Burns et al. 2010). The main advantage of digital PCR is the measurement of the absolute number of DNA copies of a transgene in comparison with the relative quantification, transgene to taxon-specific sequence, in qPCR. In addition, there is no need for a reference calibrator. The above-mentioned studies showed that dPCR is very promising, giving comparable results with qPCR, although some bias related to expectations based on predicted copy numbers was observed (Burns et al. 2010). Existing qPCR methods have to be re-evaluated to be used in dPCR and adapted if needed. Special factors need to be considered during analyses, such as the appropriate range of copies per reaction run and size of the DNA template. The method has high metrological capacities and can also be used for certifying GMO reference materials in terms of copy number ratio (Corbisier et al. 2010).

Sequencing

Sequencing is becoming more and more efficient, having higher throughput as well as becoming quicker and cheaper. Many different next-generation sequencing platforms are available and new ones are in development (http://www.dnasequencing.org/future-outlook). Therefore it is expected that in the near future sequencing can become one of the possible technologies for identifying GMOs. Sequencing can have additional advantages over other DNA-based methods, allowing detection even when the sequence of rDNA is not known, enabling the identification of unauthorized GMOs without any prior knowledge on the nucleotide composition. Results can be used not only for detection but also for risk assessment studies. A pilot study using high-throughput sequencing and computational subtraction on GM *Arabidopsis thaliana* plants has shown that rDNA can be identified from GM plants (Tengs et al. 2009). Known sequences of organisms under investigation are a prerequisite for this approach. Fortunately, there are many whole genome sequencing projects underway, allowing some optimism for wider use of the sequencing approach for GMO identification (http://www.ncbi.nlm.nih.gov/genomeprj). Sequencing may also offer new detection and identification opportunities for products developed through new techniques and for which detection and identification are presently challenging with the PCR technology. Also, use of sequencing for GMO diagnostics presents some limitations. The equipment is not affordable for all laboratories and if samples are sent for sequencing in another laboratory, the cost is still high, even though prices are decreasing each year. Moreover, sequencing needs more time to get results than if qPCR is performed on samples. In addition, next generation sequencing is producing a large amount of data that needs strong bioinformatics support (software tools and qualified staff).

Bioinformatics Support

Decision Support Systems for Smart Test Selection and Interpretation of Results

The selection of assays to be used during the testing procedure is becoming more and more complex because of the taxonomical and biotechnological diversity of numerous GMOs and the enormous amount of data on their genetic elements. Therefore, soon it will not be possible to select the assays intuitively. The so-called matrix approach was developed to allow systematic selection of genetic elements to be tested and for interpretation of experimental results.

Bioinformatics tools can provide a great deal of support to laboratory practice. Kralj Novak and collaborators have developed an algorithm that systematically searches through possible testing strategies and generates cost-effective GMO traceability testing strategies (Kralj Novak et al. 2009). Another algorithm, GMOseek, was developed within the project GMOseek, under the European ERA-NET consortium SAFEFOODERA. This algorithm shares a common function with GMOtrack to propose new screening methods that should be developed for cost-effective GMO traceability. It is also equipped with a simple decision support system, alerting the user of the potential presence of unauthorized GMOs or stacked events (Fig. 21). The main idea of these algorithms is a shift from "the same strategy for all samples" to "sample-centered GMO testing strategies." Sets of methods proposed by algorithms are selected according to a number of sample parameters (presence of plant species, possible GMOs, target elements, type of sample, frequency of previous occurrences, and a cost function). It also supports decisions in each step of the analyses as well as further evaluation of results.

The decision support system included in the GMOseek algorithm is based on the matrix approach for proposing optimum combinations of screening methods adapted to a given sample analysis. It makes use of an input GMO matrix table in which each GM event is defined as a combination of genetic elements. After loading the matrix table, the search module of the GMOseek algorithm proposes combinations of screening elements to be targeted to offer optimal GMO coverage and cost-efficiency. From this information, new singleplex or multiplex screening methods targeting some of the proposed genetic elements can be developed as a complement to the already existing screening methods. The analyst proceeds to the experimental screening phase and then indicates the results from this phase to the algorithm. The inspection module from the algorithm suggests possible GMOs present in the sample as well as the ones that should not be present. The analyst then proceeds to the experimental identification phase and indicates the results of the event-specific tests. The GMOseek algorithm compares the screening and identification results (consistency check) and suggests if possible contradictions exist, leaving the results unexplained. In that case, further investigation by the analyst is needed. If no contradiction is found, the analyst concludes using the GMO events present in the sample.

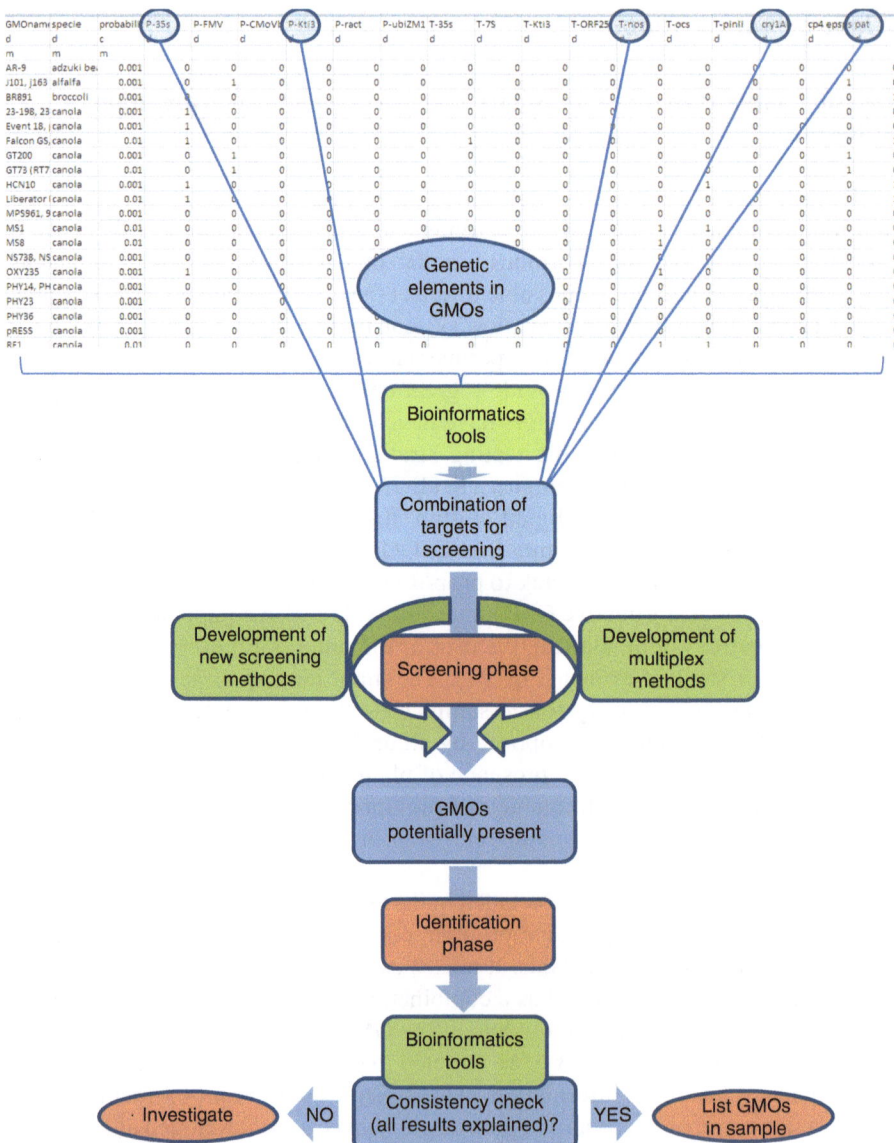

Fig. 21 Example of the prototype GMOseek algorithm. Green indicates the activities resulting from the GMOseek project (new singleplex and multiplex screening methods, bioinformatics tools). Blue indicates the inputs (GMOmatrix) and outputs (proposed combination of screening targets, GMOs potentially present in the sample, consistency check) from the bioinformatics tools. Orange indicates the experimental phases (screening and identification) and action from the analyst (investigate in case of unexplained results, list of GMOs in the sample)

The system is independent of the type of analyses; it can be for support of single-plex or multiplex PCR analyses, ready-to-use plates or microarrays. The GMOtrack is publicly available under the terms of the General Public License, and can be extended to other domains where complex testing is involved (http://kt.ijs.si/software/GMOtrack/). The upgraded version, GMOSeek algorithm, should be made publicly available soon upon completion of the project.

The "Combinatory qPCR SYBR Green screening" system (CoSYPS) is a screening platform based on SYBR Green qPCR analyses combined with a decision support system interpreting the results of analyses. The CoSYPS approach combines detection of the presence of major commodity crops (such as soya bean, maize, oilseed rape, rice and cotton) with the detection of common genetic recombinant elements (such as the P-35S or T-NOS elements) and GMO specific elements (such as herbicide-resistance genes and insect-resistance genes). The CoSYPS decision support system DSS interprets the analytical results of SYBR Green qPCR analysis based on four values: the Cq and Tm values, and the LOD, and LOQ for each method. Together, these analyses provide guidance for semiquantitative estimation of GMO presence in a food and feed product (Van den Bulcke et al. 2010).

Bioinformatics for Smart Sampling

An original documentation-based screening for products that potentially contain unauthorized GMOs was recently proposed (Ruttink et al. 2010b). This approach makes use of information and knowledge technologies. It is based on the collection of documented information on specific events, and products derived thereof. During the research, authorization, and commercialization stages including advertising, information is generated that pinpoints a GMO-derived product that can be purchased on the market. Knowledge technologies may be used to exploit the documentation space, that is, to collect, structure, and interpret documented information, and to extract knowledge from it with the purpose of discovering the unauthorized presence of GMOs on the market or in the environment. A model (a schematic computational representation of a process or system) is used to support the collection, structuring, and understanding of the information. Web search engines and specialized contextual ranking tools are used to identify the most relevant documents for each event. Advanced text-mining tools may be used to associate documents with a particular event, activity, or actor in the model, and to establish novel links among them. The idea is using Web-based search followed by manual processing of search results. Once the documentation around a "suspect" product is gathered, it can be used for experimental analytical confirmation if it is derived from a GMO and whether it is authorized for commercialization.

The efficacy of this approach was demonstrated by the discovery of unauthorized commercialized GM *Arabidopsis* in tablets intended for patients suffering from vitamin B_{12} deficiency (Ruttink et al. 2010a). This case exemplifies the need to develop such knowledge-based technology tools for the future control of traceability. This approach is also applicable to other fields of food safety.

References

Allnutt TR, Ayadi M, Berben G, et al (2010) Evaluation of different machines used to quantify genetic modification by real-time PCR. J AOAC Int 93:1243–1248.

Applied Biosystems (2001) ABI PRISM 7700 Sequence Detection System User Bulletin #2: Relative Quantification of Gene Expression. http://www3.appliedbiosystems.com/cms/groups/mcb_support/documents/generaldocuments/cms_040980.pdf. Accessed 13 May 2011.

Bahrdt C, Krech A, Wurz A, et al (2010) Validation of a newly developed hexaplex real-time PCR assay for screening for presence of GMOs in food, feed and seed. Anal Bioanal Chem 396:2103–2112.

Barampuram S, Zhang ZJ (2011) Recent Advances in Plant Transformation. In: James A.A.Birchler (ed) Plant Chromosome Engineering. 701th edn. Springer, New York Dordrecht Heidelberg London.

Bellocchi G, Berben G, Janssen E, et al (2010a) Outline of the implementation of the modular validation approach in practice and report on the validation of analytical and plug-in modules and harmonisation of quantitative data processing.; Report No.: Deliverable 4.1 of Co-Extra project.

Bellocchi G, De Giacomo M, Foti N, et al (2010b) Testing the interaction between analytical modules: an example with Roundup Ready(R) soybean line GTS 40-3-2. BMC Biotechnol 10:55.

Bhat S, Herrmann J, Armishaw P, et al (2009) Single molecule detection in nanofluidic digital array enables accurate measurement of DNA copy number. Anal Bioanal Chem 394:457–467.

Bonfini L, Moens W, Ben E, et al (2007) Analytes and related PCR primers Used for GMO detection and quantification. Luxembourg: European Communities; Report No.: EUR 23059-EN 2007.

Breyer D, Herman P, Brandenburger A, et al (2009) Commentary: Genetic modification through oligonucleotide-mediated mutagenesis. A GMO regulatory challenge? Environ Biosafety Res 8:57–64.

Buh Gasparic M, Tengs T, La Paz J, et al (2010) Comparison of nine different real-time PCR chemistries for qualitative and quantitative applications in GMO detection. Anal Bioanal Chem 396:2023–2029.

Burns MJ, Burrell AM, Foy C (2010) The applicability of digital PCR for the assessment of detection limits in GMO analysis. European Food and Research Technology 231:353–362.

Bustin SA, Benes V, Garson JA, et al (2009) The MIQE guidelines: minimum information for publication of quantitative real-time PCR experiments. Clin Chem 55:611–622.

Cankar K, Stebih D, Dreo T, et al (2006) Critical points of DNA quantification by real-time PCR--effects of DNA extraction method and sample matrix on quantification of genetically modified organisms. BMC Biotechnol 6:37.

Ciabatti I, Froiio A, Gatto F, et al (2006) In-house validation and quality control of real-time PCR methods for GMO detection: a practical approach. Dev Biol (Basel) 126:79–86.

Codex Alimentarius Commission (1997) Guidelines For The Assessment Of The Competence Of Testing Laboratories Involved In The Import And Export Control Of Food - CAC/GL 27–1997. Roma: FAO/WHO Joint Publications; Report No.: CAC/GL 27–1997.

Codex Alimentarius Commission (2004) Guidelines on measurement uncertainty - CAC/GL 54–2004. Roma: FAO/WHO Joint Publications; Report No.: CAC/GL 54–2004.

Codex Alimentarius Commission (2010) Procedure Manual. Secretariat of the Joint FAO/WHO Food Standards Programme, FAO, Rome.

Codex Committee On Methods Of Analysis And Sampling (2010) Guidelines On Performance Criteria And Validation Of Methods For Detection, Identification And Quantification Of Specific DNA Sequences And Specific Proteins In Foods. CAC/GL 74–2010. Rome: Codex alimentarius commission - WHO.

Corbisier P (2007) Application note 5: Use of Certified Reference Material for the quantification of GMO in DNA copy number ratio. Luxembourg: European Communities.

Corbisier P, Bhat S, Partis L, et al (2010) Absolute quantification of genetically modified MON810 maize (*Zea mays* L.) by digital polymerase chain reaction. Anal Bioanal Chem 396:2143–2150.

CRL-GMFF (2008) Definition of minimum Performance requirements for analytical methods of GMO testing. http://gmo-crl.jrc.ec.europa.eu/doc/Min_Perf_Requirements_Analytical_methods. pdf. Accessed 14 May 2011.

Debode F, Marien A, Janssen E, et al (2010) Design of multiplex calibrant plasmids, their use in GMO detection and the limit of their applicability for quantitative purposes owing to competition effects. Anal Bioanal Chem 396:2151–2164.

Demeke T, Jenkins G (2009) Influence of DNA extraction methods, PCR inhibitors and quantification methods on real-time PCR assay of biotechnology-derived traits. Anal Bioanal Chem .

Dobnik D, Morisset D, Gruden K (2010) NAIMA as a solution for future GMO diagnostics challenges. Anal Bioanal Chem 396:2229–2233.

Dong W, Yang L, Shen K, et al (2008) GMDD: a database of GMO detection methods. BMC Bioinformatics 9:260.

Dorries HH, Remus I, Gronewald A, et al (2010) Development of a qualitative, multiplex real-time PCR kit for screening of genetically modified organisms (GMOs). Anal Bioanal Chem 396:2043–2054.

Dymond M, Hurr K (2010) The Global Status of Commercialised Genetically Modified Plants. Wellington, New Zealand: MAF - Ministry of Agriculture and Forestry, New Zealand.

EA-EUROLAB-EURACHEM Permanent Liaison Group (2001) The Scope of Accreditation and Consideration of Methods and Criteria for the Assessment of the Scope in Testing. Paris, France: European network for accreditation; Report No.: EA-2-05.

EURACHEM (1998) The Fitness for Purpose of Analytical Methods. Teddington, UK: LGC Ltd

EURACHEM-CITAC Working Group (2000) Quantifying Uncertainty in Analytical Measurement - EURACHEM/CITAC Guide CG 4.

EURACHEM-CITAC Working Group (2007) Eurachem/CITAC Guide: Use of uncertainty information in compliance assessment. Eurachem.

EU-RL GMFF (2011) Status of dossiers. http://gmo-crl.jrc.ec.europa.eu/statusofdoss.htm. Accessed 12 September 2011.

European Commission (2003a) Regulation (EC) No 1829/2003 of the European Parliament and of the Council of 22 September 2003 on genetically modified food and feed. Off J Eur Union L 268:1–23.

European Commission (2003b) Regulation (EC) No 1830/2003 of the European Parliament and of the Council of 22 September 2003 concerning the traceability and labelling of genetically modified organisms and the traceability of food and feed products produced from genetically modified organisms and amending Directive 2001/18/EC. Off J Eur Union L 268:24–28.

European Commission (2004a) Commission regulation (EC) No 641/2004 of 6 April 2004 on detailed rules for the implementation of Regulation (EC) No 1829/2003 of the European Parliament and of the Council as regards the application for the authorisation of new genetically modified food and feed, the notification of existing products and adventitious or technically unavoidable presence of genetically modified material which has benefited from a favourable risk evaluation. Off J Eur Union L 102:14–25.

European Commission (2004b) Recommendation 2004/787/EC: Commission recommendation of 4 October 2004 on technical guidance for sampling and detection of genetically modified organisms and material produced from genetically modified organisms as or in products in the context of Regulation (EC) No 1830/2003. Off J Eur Union L 348:18–26.

European Commission (2004c) Regulation (EC) No 882/2004 of the European Parliament and Council of 29 April 2004 on official controls performed to ensure the verification of compliance with feed and food law, animal health and animal welfare rules. Off J Eur Union L165:1–141.

European Commission (2011) Commission Regulation (EU) No 619/2011 of 24 June 2011 laying down the methods of sampling and analysis for the official control of feed as regards presence of genetically modified material for which an authorisation procedure is pending or the authorisation of which has expired. Off J Eur Union L166:9–13.

European Network of GMO Laboratories (2011) Verification of real time PCR methods for GMO testing when implementing interlaboratory validated methods - Guidance document from the European Network of GMO laboratories (ENGL). Luxembourg: Publications Office of the European Union; Report No.: EUR24790 EN.

European Union Reference Laboratory for GM Food and Feed, European Network of GMO Laboratories (2010) Compendium of reference methods for GMO analysis. Luxembourg: Publications Office of the European Union; Report No.: EUR 24526 EN.

Fantozzi A, Ermolli M, Marini M, et al (2008) Innovative Application of Fluorescent Microsphere Based Assay for Multiple GMO Detection. Food Anal Methods 1:10–17.

Faye L, Gomord V (2010) Success stories in molecular farming - a brief overview. Plant Biotechnol J 8:525–528.

Gaudron T, Peters C, Boland E, et al (2009) Development of a quadruplex-real-time-PCR for screening food for genetically modified organisms. Eur Food Res Technol 229:295–305.

Golovan SP, Meidinger RG, Ajakaiye A, et al (2001) Pigs expressing salivary phytase produce low-phosphorus manure. Nat Biotech 19:741–745.

Grohmann L, Busch U, Pecoraro S, et al (2011) Collaborative trial validation of a construct-specific real-time PCR method for detection of genetically modified linseed event CDC Triffid FP967. Eur Food Res Technol :1–5.

Gruere GP, Rao SR (2007) A review of international labeling policies of genetically modified food to evaluate India's proposed rule. AgBioForum 10:51–64.

Guo J, Yang L, Chen L, et al (2011) MPIC: A High-Throughput Analytical Method for Multiple DNA Targets. Anal Chem :null.

Hohne M, Santisi C, Meyer R (2002) Real-time multiplex PCR: An accurate method for the detection and quantification of 35S-CaMV promoter in genetically modified maize-containing food. Eur Food Res Technol 215:59–64.

Holst-Jensen A (2007) Sampling, detection, identification and quantification of genetically modified organisms (GMOs). In: Pico Y (ed) Food Toxicants Analysis. Techniques, Strategies and Developments. Elsevier, Amsterdam, Netherlands.

Holst-Jensen A (2009) Testing for genetically modified organisms (GMOs): Past, present and future perspectives. Biotechnol Adv 27:1071–1082.

Holst-Jensen A, Berdal KG (2004) The modular analytical procedure and validation approach and the units of measurement for genetically modified materials in foods and feeds. J AOAC Int 87:927–936.

Holst-Jensen A, De Loose M, Van den Eede G (2006) Coherence between legal requirements and approaches for detection of genetically modified organisms (GMOs) and their derived products. J Agric Food Chem 54:2799–2809.

Hubner P, Waiblinger HU, Pietsch K, et al (2001) Validation of PCR methods for quantitation of genetically modified plants in food. J AOAC Int 84:1855–1864.

International Organization for Standardization (1994) ISO 5725:1994. Accuracy (trueness and precision) of measurement methods and results. International Organization for Standardization, Geneva, Switzerland.

International Organization for Standardization (2005a) ISO 21569:2005 Foodstuffs – Methods of analysis for the detection of genetically modified organisms and derived products – Qualitative nucleic acid based methods . International Organization for Standardization, Geneva, Switzerland.

International Organization for Standardization (2005b) ISO 21570:2005 Foodstuffs – Methods of analysis for the detection of genetically modified organisms and derived products – Quantitative nucleic acid based methods. International Organization for Standardization, Geneva, Switzerland.

International Organization for Standardization (2005c) ISO 21571:2005. Foodstuffs - Methods of analysis for the detection of genetically modified organisms and derived products - Nucleic acid extraction. International Organization for Standardization, Geneva, Switzerland.

International Organization for Standardization (2005d) ISO/IEC 17025 - General requirements for the competence of testing and calibration laboratories. International Organization for Standardization, Geneva, Switzerland.

International Organization for Standardization (2006) ISO 24276:2006. Foodstuffs - Methods of analysis for the detection of genetically modified organisms and derived products - General requirements and definitions. International Organization for Standardization, Geneva, Switzerland.

International Organization for Standardization (2007) ISO/IEC Guide 99:2007. International vocabulary of metrology - Basic and general concepts and associated terms (VIM). International Organization for Standardization, Geneva, Switzerland.

International Organization for Standardization (2010) ISO 21748:2010 - Guidance for the use of repeatability, reproducibility and trueness estimates in measurement uncertainty estimation. International Organization for Standardization, Geneva, Switzerland.

International Organization for Standardization (2011) ISO 3534–2:2006. Statistics - Vocabulary and symbols - Part 2: Applied statistics. International Organization for Standardization, Geneva, Switzerland.

James C (2011) Global status of commercialized biotech/GM crops 2010. Ithaca, NY.: The International Service for the Acquisition of Agri-biotech Applications (ISAAA).

Joint Research Centre - Institute for Health and Consumer Protection (2006) The Analysis of Food Samples for the Presence of Genetically Modified Organisms - User Manual . Office for Official Publications of the European Communities, Luxembourg.

Kodama T, Kurosawa Y, Kitta K, et al (2010) Tendency for interlaboratory precision in the GMO analysis method based on real-time PCR. J AOAC Int 93:734–749.

Kralj Novak P, Gruden K, Morisset D, et al (2009) GMOtrack: Generator of Cost-Effective GMO Testing Strategies. J AOAC Int 92:1739–1746.

Leimanis S, Hamels S, Naze F, et al (2008) Validation of the performance of a GMO multiplex screening assay based on microarray detection. Eur Food Res Technol 227:1621–1632.

Magnusson B, Naykki T, Hovind H, et al (2004) Handbook For Calculation of Measurement uncertainty In Environmental laboratories. Espoo,Finland: Nordtest; Report No.: TR 537.

Marris E (2010) Transgenic fish go large. Nature 467:259.

Morisset D, Dobnik D, Hamels S, et al (2008) NAIMA: target amplification strategy allowing quantitative on-chip detection of GMOs. Nucleic Acids Res 36:e118.

Morisset D, Demsar T, Gruden K, et al (2009) Detection of genetically modified organisms-closing the gaps. Nat Biotechnol 27:700–701.

Nadal A, Coll A, La Paz JL, et al (2006) A new PCR-CGE (size and color) method for simultaneous detection of genetically modified maize events. Electrophoresis 27:3879–3888.

Nadal A, Esteve T, Pla M (2009) Multiplex Polymerase Chain Reaction-Capillary Gel Electrophoresis: A Promising Tool for GMO Screening-Assay for Simultaneous Detection of Five Genetically Modified Cotton Events and Species. J AOAC Int 92:765–772.

Priel M (2009) From GUM to alternative methods for measurement uncertainty evaluation. Accred Qual Assur 14:235–241.

Querci M, Van den Eede G, Jermini. M, et al (2004) Analysis of Food Samples for the Presence of Genetically Modified Organisms. Luxembourg: Office for Official Publications of the European Communities, Luxembourg, Luxembourg.

Querci M, Foti N, Bogni A, et al (2009) Real-Time PCR-Based Ready-to-Use Multi-Target Analytical System for GMO Detection. Food Anal Methods 2:325–336.

Querci M, Van den Bulcke M, Žel J, et al (2010) New approaches in GMO detection. Anal Bioanal Chem 396:1991–2002.

Rommens CM, Haring MA, Swords K, et al (2007) The intragenic approach as a new extension to traditional plant breeding. Trends Plant Sci 12:397–403.

Ruttink T, Demeyer R, Van Gulck E, et al (2010a) Molecular toolbox for the identification of unknown genetically modified organisms. Anal Bioanal Chem 396:2089.

Ruttink T, Morisset D, Van Droogenbroeck B, et al (2010b) Knowledge-technology-based discovery of unauthorized genetically modified organisms. Anal Bioanal Chem 396:1951–1959.

Scholtens I, Kok E, Hougs L, et al (2010) Increased efficacy for in-house validation of real-time PCR GMO detection methods. Anal Bioanal Chem 396:2213–2227.

Shukla VK, Doyon Y, Miller JC, et al (2009) Precise genome modification in the crop species Zea mays using zinc-finger nucleases. Nature 459:437–441.

Sourrouille C, Marshall B, Lienard D, et al (2009) From Neanderthal to Nanobiotech: From Plant Potions to Pharming with Plant Factories. In: Faye L, Gomord V (eds) Recombinant proteins from plants - methods and protocols. Humana Press, New York.

Stein AJ, Rodriguez-Cerezo E (2009a) The global pipeline of new GM crops: Implications of asynchronous approval for international trade. Luxembourg: Office for Official Publications of the European Communities; Report No.: EUR 23486 EN.

Stein AJ, Rodriguez-Cerezo E (2009b) What can data on GMO field release applications in the USA tell us about the commercialisation of new GM crops? Luxembourg: Office for Official Publications of the European Communities; Report No.: JRC52545.

Stein AJ, Rodriguez-Cerezo E (2010) International trade and the global pipeline of new GM crops. Nat Biotech 28:23–25.

Taverniers I, De Loose M, Van Bockstaele E (2004) Trends in quality in the analytical laboratory. II. Analytical method validation and quality assurance. TrAC Trends in Analytical Chemistry 23:535–552.

Tengs T, Zhang H, Holst-Jensen A, et al (2009) Characterization of unknown genetic modifications using high throughput sequencing and computational subtraction. BMC Biotechnol 9:87.

Thompson M, Ellison SLR, Wood R (2002) Harmonized Guidelines for Single-Laboratory Validation of Methods of Analysis (IUPAC Technical Report). Pure Appl Chem 74:835–855.

Trapmann S (2006) Use of Certified Reference Material for the quantification of GMO in food and feed. Luxembourg: European Communities; Report No.: 4.

Trapmann S, Burns M, Broll H, et al (2009) Guidance Document on Measurement Uncertainty for GMO Testing Laboratories. v2. Luxembourg: European Communities, 2009; Report No.: Report EUR 22756 EN/2.

Trapmann S, Corbisier P, Schimmel H, et al (2010) Towards future reference systems for GM analysis. Anal Bioanal Chem 396:1969–1975.

Van den Bulcke M, De Schrijver A, de Bernardi D, et al (2007) Detection of genetically modified plant products by protein strip testing: an evaluation of real-life samples. Eur Food Res Technol 225:49–57.

Van den Bulcke M, Lievens A, Barbau-Piednoir E, et al (2010) A theoretical introduction to "Combinatory SYBRGreen qPCR Screening", a matrix-based approach for the detection of materials derived from genetically modified plants. Anal Bioanal Chem 396:2113–2123.

Vessman J, Ralucca SI, van Staden JF, et al (2001) Selectivity in analytical chemistry (IUPAC Recommendations 2001). Pure Appl Chem 73:1381–1386.

von Gotz F (2010) See what you eat-broad GMO screening with microarrays. Anal Bioanal Chem 396:1967.

Waiblinger HU, Boernsen B, Pietsch K (2008a) Praktische Anwendung für die Routineanalytik - Screening-Tabelle für den Nachweis zugelassener und nicht zugelassener gentechnisch veränderter Pflanzen/Practical application for routine analysis - screening table for the screening of authorized and unauthorized genetically modified plants. Deut Lebensm Rundschau 104:261–264.

Waiblinger HU, Ernst B, Anderson A, et al (2008b) Validation and collaborative study of a P35S and T-nos duplex real-time PCR screening method to detect genetically modified organisms in food products. Eur Food Res Technol 226:1221–1228.

Waiblinger HU, Grohmann L, Mankertz J, et al (2010) A practical approach to screen for authorised and unauthorised genetically modified plants. Anal Bioanal Chem 396:2065–2072.

Weighardt F (2006) Quantitative PCR for the detection of GMOs. In: Querci M, Jermini.M, Van den Eede G (eds) Training course on the Analysis of Food Samples for the Presence of Genetically Modified Organisms. 2006th edn. Luxembourg: Office for Official Publications of the European Communities, Luxembourg, Luxembourg.

Weitzel MLJ, Lee MS, Smoot M, et al (2007) How to Meet ISO 17025 Requirements for Method, MD, USA: AOAC International.

Žel J, Cankar K, Ravnikar M, et al (2006) Accreditation of GMO detection laboratories: Improving the reliability of GMO detection. Accred Qual Assur :1–6.

Žel J, Gruden K, Cankar K, et al (2007) Calculation of measurement uncertainty in quantitative analysis of genetically modified organisms using intermediate precision-a practical approach. J AOAC Int 90:582–586.

Žel J, Mazzara M, Savini C, et al (2008) Method Validation and Quality Management in the Flexible Scope of Accreditation: An Example of Laboratories Testing for Genetically Modified Organisms. Food Anal Methods 1:61–72.

Zhang D, Corlet A, Fouilloux S (2008a) Impact of genetic structures on haploid genome-based quantification of genetically modified DNA: theoretical considerations, experimental data in MON 810 maize kernels (*Zea mays*) and some practical applications. Transgenic Res 17:393–402.

Zhang H, Yang L, Guo J, et al (2008b) Development of One Novel Multiple-Target Plasmid for Duplex Quantitative PCR Analysis of Roundup Ready Soybean. J Agric Food Chem 56:5514–5520.

References

Index

J. Žel et al., *How to Reliably Test for GMOs*, SpringerBriefs in Food, Health, and Nutrition,   97
DOI 10.1007/978-1-4614-1390-5, © Jana Žel, Mojca Milavec, Dany Morisset,
Damien Plan, Guy Van den Eede, Kristina Gruden 2012